PRAISE FOR *THINKING WITH MAPS: UNDERSTANDING THE WORLD THROUGH SPATIALIZATION*

"This is a fascinating book! If you would like to know what Paris street names have in common with the study of poverty in nineteenth-century London, are curious about how fossilized footprints discovered in England shed light on the daily life of humans who lived 850,000 years ago, or wonder how maps can be used in the classroom to promote democratic education, this is the book for you."—**Paul Horwitz, senior scientist, Concord Consortium**

"Bruce's fascinating and insightful book draws the reader into a wide-ranging exploration of the nature of maps and their vital roles in the process of education and global understanding."—**Michael H. Fisher, Danforth Professor of History, Oberlin College, United States**

"The author connects 'maps and map-making' to virtually every discipline and human endeavor by leveraging his unusual multidisciplinary experience."—**Nama Budhathoki, executive chairman, Kathmandu Living Labs, Nepal**

"This book is a welcome addition to the field of education at all levels and disciplines. By emphasizing maps as visual language, equally important as languages of words and math symbols, it address a serious deficit in the way maps are currently taught, largely as an afterthought, relegated to 'map questions' or simple identification of place names and locations. This book, by contrast, identifies maps and other graphics as essential elements in helping us make connections across time, space, and subject material."—**Gary Benenson, project director of city technology, City College of New York**

"What emerges from Bruce's description of thinking with maps is the hugely important and often overlooked role of context—the rich network of influences and interactions that follow from broad-sense 'location.' Language and mathematics both struggle to deal effectively with context—language primarily because of its inherent linearity and mathematics because interactions and combinatorics overwhelm its methodology.

This book's novelty comes from providing the concept of visualized contextual thinking with both a strong philosophical foundation and important real-world applications. Profoundly important, in my view. We all do much of our thinking with maps. This book shows just how broadly this idea applies and how profound are its implications: for teaching, for learning, for improving society. An insightful and inspiring perspective."—**Philip H. Crowley, professor of biology, University of Kentucky**

"Bruce's point that maps are tools for thinking, in the way of language or mathematics, is significant at this time of accelerating global interconnectedness. Moreover, the global existential threats of the coronavirus pandemic and climate change make this reconsideration of maps from an interdisciplinary perspective timely. Anyone watching the evening news sees countless examples of the diverse ways that information about the progress of the pandemic is communicated in words, carefully plotted in graphs, expressed in numbers, and mapped in geographic regions. Bruce calls our attention to the perennial interest of the education community in maps and how they may be used to promote interdisciplinary education.

Bruce's breadth of experience working with students from marginalized groups in a variety of settings, including the United States, Alaska (a distinct enough part of the United States to merit separate notice), and Nepal, lends authority to his theoretical constructs. The breadth of his academic background and interests is refreshing in an age of specialists. Moreover, this broad background lends authority to his interest in promoting interdisciplinary studies.

The scope of this book makes real Bruce's claim that maps serve far greater purposes than the simple absorption of geographic information. His intellectual disposition toward inquiry and his affinity to inquiry-based learning is embodied in his presentation of maps as tools of inquiry into diverse subject matter. Although this book should be of special interest to educators, I agree with his statement: 'This book can be a resource for intelligent use of maps in any endeavor.'

Bruce explodes our traditional conception of maps as he considers maps as tools of inquiry in diverse fields. He continually reaches beyond maps to consideration of what maps afford us, even a view of ourselves."—**Sam Brian, professor of education, City University of New York**

"I recommend this book very highly—it's a read that makes one feel in the company of an intelligent mind, who is taking one's hand and leading one down a path of fascinating, interconnected thoughts on the importance and meaning of maps, which are used to gain a better appreciation of a region's physical and social history. Read this as if one has become an explorer, invited down a timeline and across the earth to learn from the revelations along the journey. From the old fold-up maps we all struggled with during car trips to the technological advances of current cartography, maps reveal culture, values, and politics, and claim territory or indeed even assert the existence of a people, such as the recent locating of thousands of indigenous communities in Peru. This is a valuable journey, spent with a well-informed traveler."—**Mary Grizzard, professor of national security affairs (retired), National Defense University, Washington, DC**

"This book is a fascinating collection of ideas about maps, their many forms, and the way people use them to both influence and be influenced by their environment. Maps and map reading are presented as personal experiences. Every map has a maker, and every map has a user."—**Abby Kerlin and Ellen McCrum, adjunct faculty, Bank Street College of Education**

Thinking with Maps

Thinking with Maps

Understanding the World through Spatialization

Bertram C. Bruce

ROWMAN & LITTLEFIELD
Lanham • Boulder • New York • London

Published by Rowman & Littlefield
An imprint of The Rowman & Littlefield Publishing Group, Inc.
4501 Forbes Boulevard, Suite 200, Lanham, Maryland 20706
www.rowman.com

6 Tinworth Street, London, SE11 5AL, United Kingdom

British Library Cataloguing in Publication Information Available

Library of Congress Cataloging-in-Publication Data

Names: Bruce, Bertram C., author.
Title: Thinking with maps : understanding the world through spatialization / Bertram C. Bruce.
Description: Lanham : Rowman & Littlefield, [2021] | Includes bibliographical references and index. | Summary: "Thinking with Maps takes readers on a journey through both traditional and modern mapping to learn how to conceive of mapping as fundamental to cognition and, thus, to what it means to be human. Each chapter considers an aspect of how we use maps. Examples from around the world show how learning can be made more relevant"—Provided by publisher.
Identifiers: LCCN 2021003826 (print) | LCCN 2021003827 (ebook) | ISBN 9781475859287 (cloth) | ISBN 9781475859294 (paperback) | ISBN 9781475859300 (epub)
Subjects: LCSH: Maps. | Thought and thinking. | Cognitive learning theory.
Classification: LCC GA151 .B78 2021 (print) | LCC GA151 (ebook) | DDC 912.01/4—dc23
LC record available at https://lccn.loc.gov/2021003826
LC ebook record available at https://lccn.loc.gov/2021003827

To Emily and Stephen, who helped me
to see the beauty of maps and
the wonderful journeys they inspire

Contents

Foreword

The perspective of this book is ambitiously holistic, a powerful and utilitarian framework that emerges from a broad view of human experience and understanding. The author is as much a Renaissance man as there is these days, blessed with a knack for expressing himself simply, clearly, and engagingly. What he says here links strongly to a scientific perspective that stays mainly beneath the surface, perhaps in the interest of remaining within everyone's comfort zone. As an ecologist, I see my duty here at the outset as underscoring the important punch of this narrative in the scientific realm.

The central theme of ecological science is the role of environmental context in the structure and function of living systems. This viewpoint crystallized with A. G. Tansley's groundbreaking concept of the ecosystem in the 1935, recognizing the integration of an ecological community with its physical and chemical environment as a fundamental ecological unit. Here the immediate context of living organisms was so central to their ability to thrive that it needed to be incorporated into an ecological meta-entity. This diverse linkage of components, organic and inorganic, dynamically intertwined, is a metaphor for integrated systems more widely, and, in particular, for the sweeping view of learning and social realization that Bertram C. Bruce developed.

The dynamics of ecological systems became more tractable when addressed locally, with an artificial boundary around a small piece of the world. But then in the 1970s and 1980s, ecologists discovered space. Populations of organisms are actually subdivided into semi-isolated subsets linked by dispersal (the jargon is "metapopulations"), often accounting for their interactions with other organisms and their basic ability to persist—or not. Ecosystems have their own context and form jigsaw puzzle tessellations with other ecosystems, resulting in landscapes. The relatively new field of urban ecology arguably arose from this landscape thinking, because the ecological viability of urban environments depends so heavily on a healthy functional relationship with their rural surroundings. Ongoing intense alterations of the natural world by the expanding human footprint, particularly through alteration or elimination of habitat, make metapopulation and landscape concepts even more ecologically central. In other words, we have to lean ever more heavily on map-based thinking about nature. Robert MacArthur anticipated this development in 1972 when he defined geographic ecology as "the search for patterns of plant and animal life that can be put on a map."

Like Bruce, I had an early fascination with maps (including the *National Geographic* globe he mentions) that led me into the study of geology. Geologic understanding is utterly dependent on maps to illustrate both the horizontal and vertical distribution of rock strata, where the primarily horizontal maps of geologic horizons can tell us about paleo-environments in space while cross-sectional maps illustrate a temporal sequence. Historical geologists must map space and time together to interpret what the rocks have to say. The fossils they contain tell a story about paleoecology that was ultimately my own entrée into ecological science.

Doing science is building and testing hypotheses about the natural world, a process that usually begins with boxes-and-arrows diagrams. These are a type of conceptual map that is usually about tentative causal relationships: my best guess about some natural phenomenon might be that situation or factor A causes B, and I might test this by manipulating A to see if B is the consistent result, relative to the case in which A is held constant. For example, I may postulate that predation risk can reduce the frequency of insect pollinator visits to flowers; I attract birds to the vicinity with birdseed (or not) to test this. So, the boxes-and-arrows map helps me recognize and design the key test, the results of which help me to build a more informed and sophisticated map of the causal relationships. At the same time, science is about networks, most essentially networks of collaborators who bring complementary expertise to the analysis of complex problems.

So, any line of thinking that depends on strong influences of context and on interactions among component parts leads to maps, which may be literal depictions of spatial distributions or may simply use space as a way to display relationships. In this book, Bertram C. Bruce shows how this concept of maps and mapping has powerful implications and applications across essentially all spheres of intellectual endeavor. This perspective has both breadth and depth well characterized here in ways that we can all put to good use.

Philip H. Crowley
Professor of Biology
University of Kentucky

Preface

There's a little joke that is told about a first-time traveler from the United States to Europe: Totally immersed in a map or guidebook describing how to find a famous cathedral, he manages to walk right into its outside wall. Living thoroughly within the map, he exposes themselves as a newbie in the territory. I confess that I have been that fool many times.

Why not pull my nose out of the book and just open my eyes to the world in front of me? I am an ardent believer in the value of ordinary experience. As John McDermott stated so well, it is "seeded with surprises and possibilities for enhancement, if we but allow it to bathe over us in its own terms."[1] Don't rely on what the guidebook writer thinks; just engage with the world as it presents itself to you.

George Bernard Shaw is frequently quoted for his saying in the preface to *Saint Joan*, "In the Middle Ages people believed that the earth was flat, for which they had at least the evidence of their senses."[2] Yes, just step outside and look; it's obvious that the world is flat. The cathedral is that large building right in front of you.

However, it is not foolish obstinacy that ties me to the map or book. As Shaw goes on to say, modern science has shown us that the obvious is often not true, that much of what we now know is "magical, improbable, extraordinary, gigantic, microscopic, heartless, or outrageous." The map of the territory offers us a different perspective, which goes beyond our immediate sense impressions: Why was this cathedral built on this particular spot? When was it constructed? Where are the hidden crypts? Who is buried within its walls?

Competing ideas: On the one hand, the world and our direct experience of it is seeded with surprises and possibilities for enhancement; on the other, our knowledge can reveal things that are magical, improbable, and extraordinary beyond the first impression. A map is the key that unlocks both of these. It helps us to find the cathedral and to make our way around its buttresses and naves. It also links us to ideas about the cathedral that we might not otherwise see or hear.

Perhaps this is why I've always loved maps. They offer the promise of finding new things, even if I actually go to very few of the ones about which I fantasize. At the same time, they help me make sense of what I do experience and relate that to the experiences of others.

When I was young, my parents must have sensed that. One Christmas present was a globe, with an innovative plastic device called a geometer. Another year it was *The Times Atlas of the World*. Yet another was a map of the constellations of stars. At other times, adventure books, for example, *Robinson Crusoe* or *The Three Musketeers*, transported me to foreign lands. It mattered little that I couldn't go to 99 percent of the places I would read about. The experience of mapping a world I couldn't see helped me to appreciate what I could. Rivers, lakes, hills all became friends to know.

I remember making papier-mâché relief maps in school. But I recall even more the use of maps in play. Like the title character in Vivian Paley's *Wally's Stories*, I might say, "The big rug is the giant's castle. The small one is Jack's house."[3] For Wally, space needed to be mapped out before the *Jack and the Beanstalk* game could get underway. Nearly every activity I might engage in called for defining a space, setting its components and boundaries. That might be an enactment of a medieval jousting tournament or a Wild West adventure.

Board games always provided a map in advance, sometimes only iconically as in *Monopoly*, sometimes more realistically as in the later, computer-based *Oregon Trail*. But maps of various sorts appeared everywhere. I remember enjoying the Six Flags amusement park. The original one was not so much about thrilling rides as the parks are today. It actually made an effort to create believable versions of the six cultures that those flags represented. Exploring the park meant following a map laid out with signs and décor taking the visitor through an exploration of other times and places.

The zoo was also a map one could experience. One might walk through a South American rain forest or an African savannah. Like the parks and games, the experience was manufactured and limited, but it opened the door for the imagination.

My interest in maps led me to become more interested in the city, how it was organized and constructed. In 1959, Victor Gruen was hired to create a plan to solve the traffic problems of downtown Fort Worth, the city where I lived. He proposed banning all cars from the downtown to provide a better pedestrian experience and to revitalize the area. This was a radical idea for

Texas, especially in the 1950s. Large parking garages on the periphery were to result in no more than a two-and-a-half-minute walk to anywhere in the business district. All deliveries would be made by way of underground tunnels. Traffic would circulate around the downtown by a ring highway.

Ultimately Gruen's plan failed. Only the ring highway of his vision came to be. Gruen and other visionaries, even my middle-school class, liked his idea. But others wanted to draw a different map. I learned that maps reflect power differences. The map of those in power became the future for Fort Worth. I learned that a map is neither a completely true depiction of present or past reality, nor a sure guide to the future.

In spite of that, I was fascinated by the maps created for the Gruen plan and later considered urban planning for graduate school. To this day, I'm always interested in how towns are organized and happier still when I can see how their design relates to rivers, oceans, mountains, and other natural features, then how all of that, in turn, affects plants and wildlife.

Maps for me represent the essence of interdisciplinary or holistic understanding of the world. The fact that they are created by so many different people, for different purposes, makes them not only a challenge to understand, but also an unending source of pleasure.

"Map" is a noun, representing a thing, which holds useful information. But it can also be a verb, something we do. Maps are an opportunity to make our own mark on the world. Creating maps, or mapping, can be a powerful aid to memory and thinking.

Like T. S. Spivet, the ten-year-old cartographer in Reif Larsen's novel, we can depict any of our activities and seek thereby to make sense of them. T. S. says, "Something about measuring the distance between *here* and *there* cast off the mystery of what lay between, and since I was a child with limited empirical evidence, the unknown of what might just lie between *here* and *there* could be terrifying. I, like most children, have never been *there*. I had barely even been *here*."[4] So, he makes a map displaying "here," "there," and "terra incognito," the distance between here and there.

As T. S. shows in his maps, they can be of many forms; none of his hundred or so maps look like conventional road maps. Maps can be diagrams, charts, musical scores, Polynesian stick charts, tattoos, and more. In writing this book I hoped to continue my exploration of the world by continuing my exploration of maps. Thinking with maps—indeed, about, through, and because of maps—can be one of the most important learning experiences we can give to schoolchildren, scholars, and ourselves.

References to John Dewey's writing are to *The Collected Works of John Dewey, 1882–1953*, edited by Jo Ann Boydston (Carbondale and

Edwardsville: Southern Illinois University Press, 1969–1991). SIU Press publishes this in three series: *The Early Works* (EW), *The Middle Works* (MW), and *The Later Works* (LW). For example, LW 13:29 refers to *The Later Works*, volume 13, page 29.

NOTES

1. John Dewey, *The Philosophy of John Dewey*, edited by John J. McDermott (Chicago: University of Chicago Press, 1981), x.
2. George Bernard Shaw, *Saint Joan: A Chronicle Play in Six Scenes and an Epilogue* (London: Penguin Classics, 2001), preface.
3. Vivian Gussin Paley, *Wally's Stories: Conversations in the Kindergarten: Rulers* (Cambridge, MA: Harvard University Press, 1981), 13.
4. Reif Larsen, *The Selected Works of T. S. Spivet* (New York: Penguin, 2009), 32.3.

Acknowledgments

Many people contributed to my thinking about maps and to the text itself. Visits to Bank Street College of Education in New York were inspirational, as has been published work from there. I would like to single out specific help from Ronnie Sampson in learning about the school and its work with maps. Brett Dion added more and also worked diligently to assist with images from the archives. Salvatore Vascellaro was a helpful guide in learning about the "long trips." Abby Kerlin and Ellen McCrum helped in productive discussions, offered great examples, and commented on issues of power in mapmaking, as well as on the personal interpretation and creation of maps.

Denny O'Connell and Bob Jackson provided invaluable insights on the Wellfleet Conservation Trust and its education programs. Mike Fisher provided a very helpful reading and good suggestions. Mary Grizzard, Paul Horwitz, and Wally Feinberg read and commented in useful ways as well.

Sam Brian introduced me to his terrain maps and his important work in Nepal. Gary Benenson reminded me of the broad view of maps, which I endorse, and which he lays out so well in his book with James Neujahr. Nama Budkathoki and his team at Kathmandu Living Labs was invaluable in showing how maps can be devoted to social good. Phil Crowley wrote the foreword to the book; even more important, he helped me think through issues around visualization and spatialization, especially the role of context.

As always, my family has been very supportive, not only through encouragement and commiseration, but specifically through comments on the text and discussions of the theory. Emily and Stephen both helped me move from a set of assorted ideas to what I hope is a coherent story. My wife, Susan Bruce, has been a continual motivator, critic, listener, and all-around aide in making this happen.

Introduction

> Your political globe is much more than a masterpiece of the cartographer's art. It also represents one of mankind's greatest achievements, the exploration of our planet. . . . The sum of all they have learned up to this time is incorporated in your new Globe. Here at your fingertips are centuries of geographical knowledge accumulated at the cost of countless lives, innumerable explorations, and experiments of every conceivable nature.
>
> —Gilbert M. Grosvenor, editor, *National Geographic*[1]

Where do we begin? A treasure map, at least of the kind one might find in a children's adventure story, will tell us: go to the place marked "Start" and proceed from there. You will eventually find a buried chest of gold and jewels. The map is your guide. Even one crudely drawn can be the key for finding your way and for understanding how all the pieces of the journey fit together.

The journey here will not lead to pirated gold, but it reveals more valuable treasures. Seeking an answer to the question of what maps are and why they matter, it touches on wondrous realms of cartography from ancient marks on a cave wall to a hyperlinked, multidimensional, multimedia space for interaction. We start with an antique map, actually a globe, which might be found in an attic or at a flea market.

THE *NATIONAL GEOGRAPHIC* POLITICAL GLOBE

The introduction to the guide for the *National Geographic* political globe from the mid-twentieth century (excerpted earlier) sounds both extravagant and antiquated. Does it truly represent "one of mankind's greatest achievements

. . . the sum of all they have learned up to this time?" Does it contain "centuries of geographical knowledge accumulated at the cost of countless lives?" Questions like these might arise easily for a contemporary generation who attend primary school without a globe in every classroom.

For their parents, perhaps their grandparents, a globe was always there. It may have served as a decoration, an arcane object marking the space as a site for learning. Some saw it as needless clutter. Many learned that it was fun to spin the globe as fast as possible. Very few saw it as a "masterpiece" answering questions of "every conceivable nature." Even fewer knew that the frontispiece images showing "ideal learning" in many Enlightenment periodicals often included a globe.

The "world of learning" is both a metaphor for knowledge and a method for expanding knowledge, but that idea was lost on many schoolchildren, even in otherwise excellent schools, and most likely on their parents and teachers as well.

Global understanding is more necessary than ever in an age with new modes of communication and transportation, global trade, massive migrations, and global challenges. It seems perverse that it is the present moment when we seem to have lost both globes and our ability to use them. More generally, we have lost maps, or at least an understanding of how maps both amplify and define our being.

MAPS AS BOTH OBJECTS AND WAYS OF THINKING

The *National Geographic* political globe is undoubtedly a thing. It has dimensions and mass. It has accessories made of plastic and an accompanying paper guide. As an object it is beautiful and a marvel. Many people who encounter it would be curious about how it could be made and how they might use it.

The globe also indexes a way of thinking. It suggests first that we think of the world as a sphere, something that most of us know but rarely reflect upon.[2] It asks us to be aware of political boundaries and how they relate to rivers, lakes, mountains, islands, peninsulas, and oceans.

The globe was accompanied by the *geometer*. This was a unique, clear plastic device, including a time zone ring, that could be positioned over a location on the globe and used to calculate latitude, longitude, area, distance, great circle routes, relative times, sunrise and sunset, and other properties. The geometer was an overlay that amplified the usefulness of the globe, assisting the user to become skilled in geometry, hence the name.

A closer look reveals that the accessory geometer shows the *analemma*, a misshapen figure 8. The analemma marks the position of the Sun as seen from a fixed location on Earth at the same mean solar time for any date in the course of a year. Using the analemma encourages us to think of the Earth as a component of the solar system, and beyond.

Continuing along this line of inquiry we ask, what exactly is a map, beyond a surface description of it as a physical object? Why does it matter? How does it enter into our ways of thinking, problem solving, and appreciating the world? How might we conceive of mapping as fundamental to cognition, like literacy and mathematics?

Our exploration of these questions is important for an understanding of human language, thought, and reasoning. It can also make us more effective mapmakers and map users. Accordingly, it has major implications for teaching and learning.

By the third century BCE, Greek astronomers had established that the Earth was roughly a sphere. It was then that they produced the earliest terrestrial globes. Unfortunately, those globes have not survived. The oldest remaining one is probably the *Erdapfel*, built by Martin Behaim in Nuremberg around 1492 (figure 0.1).[3]

Figure 0.1. Children next to the oldest remaining globe, the Erdapfel, built by Martin Behaim in Nuremberg in about 1492. Ernst G. Ravenstein, *Martin Behaim, His Life and His Globe* (London, UK: George Philip & Son, 1908), 78. *Public domain*

Behaim's globe is twenty-one inches in diameter, about twice the size of the *National Geographic* globe. It is a laminated linen ball in two halves, reinforced with wood. Using papier-mâché and coated with gypsum, Georg Glockendon painted map components on parchment according to Behaim's specifications. The parchment strips include place names, pictures, banners, coats of arms, and notes about foreign monsters and countries. Some talk about travelers, for example, Marco Polo. The completed drawings were then pasted into position around the sphere.

Erdapfel is German for "earth apple." According to the Bible, God commanded Adam: "Of every tree of the garden thou mayest freely eat. But of the tree of the knowledge of good and evil, thou shalt not eat of it." Many believe that the forbidden fruit in the Garden of Eden was an apple tree. Taking a bite of one of those apples opened the door to knowledge. So, it is very appropriate that the earth apple and its descendants continue to represent the world of knowledge as well as the world we ride upon.

One could spend hours studying the globe and geometer, plus its index and guide, and never exhaust all of the possibilities. It is a minor ecosystem for learning. Even today, it portrays the world in ways that are difficult to match with the best digital interactive maps, or even virtual reality. Of course, new digital media offer their own affordances that the 1976 globe cannot match.

Despite the wonders of both material-based and digital maps, many people think of the very idea of maps as a relic from the past. They have a shoebox with old road maps, which needs to be sent to paper recycling someday. They may have a decorative historical map, framed on the wall, one that is never used as an actual map. When they need to go somewhere, they enter an address using voice input or auto-complete into the audio-enhanced GPS on their phone. It then tells them where to turn ("take exit 12 toward Orleans") or how to find a place to eat along the way. The technology is scarcely a working map, more just a pictorial background for a narrow inquiry.

LOST CONNECTIONS

Many believe that they can survive today without knowledge of or direct use of maps. We rarely think about the extensive use of maps by airplane pilots, construction crews, forestry managers, biologists, town planners, realtors, historians, archaeologists, and many others. But in severing our direct connection to maps, we hamper our ability to solve many problems; we miss out on the aesthetic appreciation of maps; and we simply lose the fun that maps can provide.

The problem is manyfold. Like words and numbers, maps help us navigate the world. In many cases, maps include text and quantification as well. Maps are vital for the ecosystem of learning.[4] To fully benefit from maps, we need to appreciate the following:

- The incredible variety and usefulness of maps across domains: literature, science, history, sports, arts, music, language, and much more.
- The diversity of forms of maps, encompassing paper-based cartographic maps, and also digital maps with layers, timelines, charts, diagrams, pictorial representations, and more.

- The potential for a richly detailed network of maps connecting them with one another and with other symbolic systems.
- How maps provide context for individual facts. A network of maps has emergent properties that provide much more than the answer to "where do I turn?"
- Maps as something constructed by other humans, expressing a point of view, a purpose, and rhetorical strategies.
- The ability to create maps, including ranging from pinning a bookmark or a walking route on an existing digital map to complex and imaginative constructions.
- Maps are always interpreted based on the user's background and purpose in various ways. They help us understand and portray experience. They expand our capacity to think and participate in the world.
- Many people have limited map literacy. They cannot read them or construct them any more than an English-only speaker can read or write Chinese characters.

GEOGRAPHY

A recent report says that although we have inspiring examples of highly effective geography education, "the amount of geography instruction that the overwhelming majority of students receive, the preparation of their teachers to teach geography, and the quality of their instructional materials are inadequate to prepare students for the demands of the modern world."[5] This may be both cause and effect of an impaired *geographic imagination.*

How might a geographic imagination impinge on simple decisions in daily life? Fran Martin, a researcher in the field of primary-level education, writes that her choice regarding drinking a cup of coffee in a café is based on such questions as the following:

> Where is the nearest café which serves the type of coffee I like best in the time I have available? Once in the café, where shall I sit so that I am warm enough, where I am least disturbed by movement in and through the café, and where I am able to do some people watching? Underlying some of these choices is an implicit understanding of place: the type of environment I prefer when drinking a cup of coffee and micro-climates: If I sit near the door there might be a draught . . . [there is] a mixture of social, economic, environmental, and political factors, all linked to places that have local dimensions (the café itself, the city I am in) and global dimensions (where the coffee is grown, for which company).[6]

A central aspect of Martin's decision making about cafés is that she cannot simply look up café on a map. It would require a highly intelligent app to satisfy her demand for a place where she will not be disturbed by movement of others but can still engage in people watching. Her additional constraints make the task undoable. Even for her, the task is formidable, requiring her to draw upon a complex, nuanced, but ill-specified set of local and global factors.

If she were to ask an online map program to display nearby cafés, she might see a cluster of waymarks identifying places that serve coffee. A fancy app could conceivably narrow the search to the type of coffee she prefers at a time she has available. But to succeed at even this ordinary task, she must engage the map with her geographic imagination.

When one addresses such issues as the current climate crisis, a limited geographic imagination has more dire consequences. Scientists with extensive knowledge of ocean and air currents and layers, mountains, rivers, lakes, forests, and deserts face a difficult task. They must also use detailed weather models, visualization tools, mathematical sophistication, and computer resources to understand the forces of climate change. Generally favoring conservative projections, they are often surprised to discover that the effects of global heating are more severe than they had expected.

For the ordinary citizen, this can be an even greater challenge. It is too much to expect that their geography knowledge meets that challenge. However, when students employ limited geographic thinking and imagination, they are ill-equipped even to understand the issues at a cursory level.

PROBLEMS IN THE WORLD TODAY

Major problems beset the world today, including the human-caused climate disruption, ecological collapse, food security, unsustainable consumption, pollution, disease, gender inequality, radical nationalism and sectarianism, authoritarianism, the concentration of wealth in the hands of a few, poverty, and the threat of nuclear war. Even a casual reflection on any of these shows that remedies are difficult to find. Moreover, those hypothesized remedies are unattainable without an informed, critically engaged population able to discuss complex issues and carefully analyze evidence. At the foundation is an ability to *see* the world, to know it beyond its immediate, local, and surface appearance.

Education by some means, formal or informal, self-directed, online, or whatever, is thus essential for addressing any of these and other global issues. Although education is, of course, far from sufficient, it is nevertheless necessary. This makes curricula or what we teach, and how we teach it, at the heart of meaningful change, and even survival. Theorizing about knowledge cannot be relegated to philosophers, nor curricula left for education experts. It must, instead, be central to any discussion aimed at preserving a livable world.

KNOWING THE WORLD THROUGH SPATIAL RELATIONS

Across many fields, we now have the tools to acquire more and better data, but how we make data become information, knowledge, or ideally some sort of wisdom is a complex process. Systems of symbols are necessary for engaging in that process. Through those systems, we define terms, create categories, and develop relationships among bits of knowledge to discern larger patterns and find solutions.

Words and grammar are the building blocks of one major system of symbols. Written language allows us to express ideas and construct theories to move us far beyond simple bits of data or information. We use it to sort and classify ideas. This work shows up in essays, laws, outlines, and keywords. At other times, we emphasize stories, exploring temporal connections through literature or in such disciplines as history and journalism.

The patterns explored through mathematics constitute a second major system of symbols. Functions reveal connections across data. Ideas of quantity and structure can be used to elaborate ideas just as textual language does. Mathematical expressions of change are an alternate way to examine temporal connections. It is no accident that these two major symbol systems are intrinsic to formal education, represented in the popular imagination through reading and writing for the first and arithmetic for the second.

As the world becomes more interconnected, and as our understanding of ecological, political, and cultural interconnections grows, understanding spatial relations and reasoning through visual media becomes more essential. Spatializations are a third system of symbols, essential for understanding existing ideas and for generating new ideas. And yet, of the three ways of organizing knowledge, the third may be the most neglected and least understood.

Moreover, the spatial, which promises connection across wide areas, is itself ironically often not connected to other areas of knowledge. The idea of spatial reasoning from mathematics is rarely linked to the spatial thinking needed for political science; the use of new digital mapping tools is still spotty and poorly integrated with other media; geography is often reduced to learning such geographic facts as place names, separated from broader understanding of relationships between social and physical worlds, or of movements through time. Mapping helps us to understand, but we also need to understand maps themselves.

Thinking with Maps: Understanding the World through Spatialization addresses this problem, developing its argument through historical analysis and cross-disciplinary examples involving maps. The idea of *maps* here includes traditional cartographic representations of physical environments but more broadly encompasses the wide variety of ways that spatializations are used across all disciplines to enable understanding, to generate new knowledge, and to effect change.[7]

The idea of *thinking* with maps is also used broadly. Maps become, not simply one among many items to learn about, but indispensable tools for thinking across every field of inquiry, in a way similar to that of textual and mathematical language. Effective use of maps becomes a way to make new knowledge, much as writing or mathematical exploration not only displays ideas, but also creates them.

Maps represent an indispensable symbol system akin to but not subsumed by language or math. They are not just a means to improve geographic knowledge, as valuable as that may be. Instead, they provide mechanisms for rejuvenating our engagement with the world, helping us to become more capable of facing our global challenges. The renewal of democracy, the value of community, and a renewal of evidence-based reasoning are crucial for society's survival. This book offers a fresh approach to these goals, highlighting the understanding and use of maps as an overlooked aspect of that renewal. It has developed with that larger public in mind.

CONSEQUENCES FOR EDUCATION

It is no accident that geography has been conceived by many as a unifying subject for education. Once freed from the shackles of rote memorization of place names, it is a topic that naturally integrates the physical and the social worlds. John Dewey wrote about this in one of his earliest books on education: "All studies arise from aspects of the one earth and the one life lived upon it. We do not have a series of stratified earths, one of which is mathematical, another physical, another historical, and so on. . . . All studies grow out of relations in the one great common world."[8]

Writers of Dewey's time and today have argued for holistic or integrated learning, inspired by this "one great common world" perspective.[9] Dewey continues, "When the child lives in varied but concrete and active relationship to this common world, his studies are naturally unified. It will no longer be a problem to correlate studies. . . . Relate the school to life, and all studies are of necessity correlated."[10] Mapping is by no means a guarantee for relating school to life, but it provides a vital perspective and set of tools for achieving that connection.

The tools and components of learning through maps pervade discussions about curriculum reform. Spatial reasoning is central to mathematical thinking.[11] Student interest in interactive virtual experiences can enable a bridge to social sciences.[12] K–12 classroom teachers, future teachers, and educators in informal settings all have an interest in promoting integrated curricula. Fluency with the new digital tools for spatial understanding, for instance, interactive maps,[13] GPS, GIS, mobile devices, and virtual reality, has become a core skill.

The most significant message of this book is not about maps per se, even when we adopt a broad definition. Instead, it is a call to open up the ways that we think about learning, and both formal and informal education. Used well, maps can become a way to support the learner's engagement with a complex and changing world, thereby extending education far beyond its sometime manifestation as a process of transmitting decontextualized knowledge to students. Geographic understanding and reasoning are central to all learning.

This book has an even broader aim: it is fundamentally about general principles of how we learn and know. It calls for a renewed focus on democratic education in which both the means and ends are democratic. Education, just as the political realm, should follow Dewey's dictum that "democratic ends need democratic methods for their realization."[14]

A JOURNEY THROUGH THE WORLD OF MAPS

The chapters here take us on a journey from a Start to "The Map Ahead." As our journey takes us through the moors and mountains of both traditional and modern mapping, we may see many marvels, but more important, we can learn how to conceive of mapping as fundamental to cognition and, thus, to what it means to be human. Each chapter considers an aspect of how we use maps.

Maps are, first of all, useful tools to help us locate ourselves and to traverse the places we find ourselves in, but more significantly, they are the means we employ to make sense of our life in context. These ideas are developed in chapter 1 ("Way Finding").

Could an entire city become a map for a country? (See chapter 2, "Constructing.") What would it be like for the city to go about its ordinary business and tell a story about itself and its history? Considering these questions gives us insights into how maps are more than repositories of facts; they are how we construct the world around us. Through these constructions, maps provide useful information, guidance, enlightenment, amusement, or indoctrination. They can also invite users to interact and to make their own maps to give meaning. These examples show how the space we live in is not pre-given but constructed through our decisions and actions.

We have an incredible variety of maps and ways of employing them for solving problems and learning. These maps are more than vital tools for diverse practical purposes and academic disciplines. They are tools for telling the story of our exploration of the world. Along with verbal language and mathematics, they create the story of a scientific inquiry and communicate it to others. Maps allow us to explore data and phenomena in new ways. They are a vital yet underappreciated tool in the scientific process across every discipline. Chapter 3 ("Learning") offers just an appetizer of these many uses of maps as learning tools.

Chapter 4 ("Expressing") considers just a few examples of the role that maps play in creative expression, visual arts and music, free exploration, and play. Each example represents a way that maps support diverse learning styles. Through maps as art our ability to imagine the physical world can grow, as well as our sensitivity to the perspectives and situations of other people. The use of maps in games, literature, and other areas can help us appreciate the importance of physical space in all aspects of our lives.

Once maps are brought together through networks, something remarkable happens. Data represented in one map comes together with data in other maps to yield information that could not be easily predicted from examination of the individual maps. Maps can be incorporated into the inquiry needed for problems of public health and other social services, into public administration, to promote cultural understanding, and a variety of other fields. They are especially effective when detailed knowledge of complex phenomena is needed. Chapter 5 ("Solving") shows a few of the ways that maps help us address problems in the world and occasionally change things for better.

Chapter 6 ("Connecting") focuses on how maps can be created and used to connect complex ideas to one another. This is especially important in the effort to promote civic intelligence. One example involves the use of an array of maps for formal education, in particular, a U.S. social studies curriculum. Following that are examples of the use of new digital tools for civic projects and promotion of civic intelligence.

Mapping does not fit neatly in the traditional curriculum, although it overlaps with every other area. As Gary Benenson and James L. Neujahr say in their book on mapping, "Nearly every topic becomes easier to understand when graphic devices are used. Imagine trying to understand history without timelines, music without scores, or geography without maps."[15] Chapter 7

("Entering the Map") explores how not only maps, but *mapping*, the process of creating maps, can be a vital aspect of learning.

Chapter 8 ("What Is a Map?") steps back to explore the very definition of maps. In the process it considers how maps are tools for thinking and learning in every imaginable area. It considers some standard definitions, the difficulty of defining such key terms as *spatial*, while seeking a broader view to encompass the diverse forms of maps today. Along the way, it discusses the organization of ideas by concept, time, and space. It emphasizes the many kinds of maps and their diverse forms and uses. It includes concept maps and how location maps interact with other ways of organizing knowledge, for example, hierarchies, or networks of concepts or time.

The conclusion proposes "The Map Ahead," taking a new perspective on maps and mapping into all of education.

THINKING WITH MAPS

It is tempting to say that we need to dust off the globes and bring the crinkly maps out of the drawer. But the problem is not with maps. They have always been available and are today being employed in every human endeavor, availing themselves of the new technologies of digital maps, overlays, audio, and virtual reality. Instead, the problem is with how we view maps and mapping.

All too often, maps are an afterthought, perhaps to be applied as an illustration for an article or as a mostly dispensable tool for modern travel. We need to recognize maps as a vital tool for understanding items that are *spatially distributed* or can be viewed as such. We need to realize that thinking with maps is as essential as reading or mathematics for human participation in the world.

NOTES

1. The globe was first produced in 1961 and continued for many years. The opening quote here is from "The National Geographic Globe," in *National Geographic Political Globe: Index and Guide* (Washington, DC: National Geographic Society, 1976).

2. More accurately, the Earth is a collection of mass that changes shape because of tidal movements, shifting of tectonic plates, and other factors. Nearly a perfect sphere, it bulges at the equator and flattens at the poles, making an oblate spheroid. It has a slight pear shape with a larger southern hemisphere surface area. It also has mountain ranges and empty plains, coral reefs, and huge underwater chasms that affect its shape. But for most purposes, sphere is a good simplification.

3. Ernst G. Ravenstein, *Martin Behaim, His Life and His Globe* (London: George Philip & Son, 1908).

4. Bertram C. Bruce, *Education's Ecosystems: Learning through Life* (Lanham, MD: Rowman & Littlefield, 2020).

5. Sarah Bednarz, Susan Heffron, and Niem Tu Huynh, eds., *A Road Map for 21st Century Geography Education: Geography Education Research* (Washington, DC: Association of American Geographers, 2012).

6. Fran Martin, *Teaching Geography in Primary Schools: Learning to Live in the World* (Cambridge, UK: Chris Kington, 2006).

7. Visualization is usually restricted to literal seeing, for example, by a sighted person. Spatialization is a broader term encompassing the process of understanding things and describing them as if distributed in space, even if they are not strictly visible.

8. John Dewey, "The School and Society," MW 1:54.

9. Susan M. Drake and Rebecca C. Burns, *Meeting Standards through Integrated Curriculum* (Alexandria, VA: ASCD, 2004); Mary Taylor Huber and Pat Hutchings, "Integrative Learning: Mapping the Terrain" (Washington, DC: Association of American Colleges and Universities, January 1, 2005).

10. Dewey, "The School and Society," MW 1:54–55.

11. National Research Council, *Learning to Think Spatially* (Washington, DC: National Academies Press, 2006).

12. Diana Stuart Sinton and Jennifer Lund, *Understanding Place: Mapping across the Curriculum* (Redlands, CA: ESRI Press, 2007).

13. For example, https://www.arcgis.com.

14. John Dewey, "Democratic Ends Need Democratic Methods for Their Realization," LW 14:367.

15. Dewey, "Democratic Ends Need Democratic Methods for Their Realization," 367.

1

Way Finding

> In May 2013, a spring storm uncovered the oldest traces of human passage outside Africa on Happisburgh's claggy foreshore. Heaving seas had stripped away the sand behind the dilapidated postwar flood defenses to reveal a section of laminated silt flecked with dozens of lozenge-shaped hollows. The scooped depressions were the 850,000-year-old fossil footprints left by a group of early humans living along the muddy banks of an ancient river.
>
> —David Farrier, *Footprints: In Search of Future Fossils*[1]

Way finding is the most obvious use for a map. You want to know not only where you are, but also how to get somewhere else. A map is a convenient visual display for that purpose. But as ancient footprints in fossilized silt illustrate, that function can be accomplished in diverse ways. A map also tells a story, which the listener can enter into. This can be the seed for all sorts of learning.

FOOTPRINTS AS MAPS

Footprints and other marks of daily life, now preserved in rock, invite us to imagine people of long ago. The trace fossils in Happisburgh, England, as Farrier describes, tell us vivid stories about how people lived, their weight, gait, and habits, extending what we can learn of their physiology and anatomy from fossilized bones and teeth.

The silt in Happisburgh resembles a step map on a dance floor. The dance suggests adults stopping to encourage tired children. They may look around to check for predators or locate food and water. From time to time they raise arms to indicate points of interest or offer a hand to others.[2] Farrier says that some impressions showed the outlines of individual toes. We must rely on his account. The footprints from 850,000 years ago had lasted long enough to provide a glimpse of life from that forgotten time, but they were washed away by the tide within two weeks.

These prints were made by Homo antecessor, the oldest example of the genus in Europe. A recent analysis of dental proteins provides evidence that it was more closely related to modern humans than to Homo erectus. The body proportions fall within the range of variation for modern humans, and upper limb proportions were likely more similar to those of modern humans than Neanderthals. The presence of large animal carcasses at settlement sites implies that they were carried to the site intact, by multimember hunting parties. They had to delay eating after a kill to share it with other group members. Coupling this evidence with that of the Happisburgh trace fossils shows social cooperation, division of labor, and food sharing.

The Happisburgh step map is the best map, the best story, we have of life from that time. It tells us where people were walking, how many there were, their ages, and a hint of their activities. We can follow along their route, thereby sharing their experiences. As they were finding their way, we find it as well. They neither wrote nor painted, at least in any way that has lasted to our time. But their steps record a story. As Rebecca Solnit points out, "one aspect of the history of walking is the history of thinking made concrete—for the motions of the mind cannot be traced, but those of the feet can."[3]

Tracing the feet of the early Happisburghians is a fascinating story per se. There is surely much more to learn about those people from nearly a million years ago. Their story also invites us to reflect on our activities and their origins. Through their step map we can enter their world and come to understand more about our own.

UNTRAVELED ROADS

Perhaps the most familiar map of all is the common road map. Those ordinary maps that are so easy to unfold and difficult to refold can be used to find our way, but like the Happisburgh footprints, they can also tell us much more than just how to get from A to B.

The same can be said of the maps embedded in smartphone apps. Their frequent use as simple direction finder obscures what they can tell us about our environment. Without at all discounting the value of maps for keeping us from getting lost, we ought to consider what else they can teach us.

Road Maps

The ordinary road map can inform the user about the size and shape of cities, the course of rivers, the location of state parks and nature areas, historic and cultural sites, and in general provide an overview of the territory. This can be useful for assessing how far one has traveled or needs to go. They help in finding the out-of-the-ordinary route or stop along the way. The map is an entrée into understanding the neighborhood or region and how everything fits together.

The United States has a huge number of currently published road maps (maps printed on paper), from town street plans to counties and small regions, to states and atlases covering the entire country. The American Automobile Association alone publishes and regularly revises more than a thousand regional, state, and local maps, including the *AAA U.S. Road Atlas*. Route maps are also available.

The need for good maps is evident when one considers the road mileage to represent. As just one example, Los Angeles has 314 miles of Interstates, many intermediate roads, and 13,319 miles of local roads for a total of 21,826 maintained road miles. There are tens of thousands of exit and entrance ramps and road junctions.[4]

Road Atlases

State tourist offices, including welcome centers located on main routes just inside state borders, provide free official state highway maps. Free state maps are also available from AAA, local tourist bureaus, and some service stations. Yet, despite the wide availability of free maps and competition from digital maps, sales of road atlases have actually been increasing.[5]

People interact with road maps more than ever before due to their availability on multiple platforms, including foldout print, books, websites, phone apps, dedicated GPS, and others. The line between online digital maps and printed maps is blurred, because many maps are available in pdf form and can be printed out inexpensively in full color at various sizes. Hybrid or linked forms are growing, including putting points of interest from an atlas into GPS systems and URLs into an atlas to get additional, especially up-to-date information online.

The various media for road maps survive and prosper because of the different purposes that users have. Some want street-level routing, and for that they depend heavily on GPS-driven maps on tablets, phones, or vehicle GPS screens. These tools have many additional advantages: they quickly show the quickest route, especially in light of current road conditions; they show your current location; they provide real-time satellite weather data and current traffic conditions; they can show layers of services and attractions; and more.

Nevertheless, many still find a printed folding map or a road atlas to be essential. They argue that the path is more important than the destination. Seeking adventure, discovery, and learning along the way, they want a resource that fosters those experiences. They contend that a printed map or atlas enables them to better get an overview;[6] see at a glance various options for routes, including areas that are likely to have attractive places to stop for meals; plan a long trip; maintain independence from batteries and Wi-Fi or cellular signals.

Either format can be a valuable resource for learning.[7] They make it possible to explore the world around us, both by showing routes for exploration and by detailing all sorts of information about the regions one cannot visit directly.

Goode's World Atlas has been a leading educational atlas since 1923. It includes general physical and political maps as well as thematic maps, graphs, and tables. It introduces itself by claiming that a map is the "most effective and realistic medium for gaining knowledge of relative location, size, direction, distance, and other facts which are related to and distributed over the earth." Further, an atlas "assembles various reference and thematic maps as well as graphic and tabular information. . . . It is a necessity for understanding the course of world events."[8]

Despite the claims of the Goode's producers, popular discourse about educational materials often ignores atlases, as well as maps, globes, digital media, and related materials. They are often viewed as obsolete relics, at best incidental to the educational process. Many would not consider maps as the most effective and realistic medium for gaining knowledge of spatially distributed facts, nor the atlas as a necessity for understanding the course of world events.

CONSERVATION LAND AND TRAILS

School curricula have been designed to reinvigorate the social contract by teaching young people about the history of their country, its geography, its demographics, and its problems. The best curricula employ a complex network of place maps, timelines, charts, images, texts, and other materials. But that learning can occur through informal education as well.

Wellfleet Trails

An example of a network of maps for citizens in general can be seen in the town of Wellfleet, Massachusetts. Various organizations provide opportunities for the public to enjoy nature observation, photography, exercise, dog walking, meditation, and other benefits of the town's valuable natural resources, particularly by taking advantage of several contiguous protected properties.[9] See figure 1.1 for one simple map within the National Seashore.

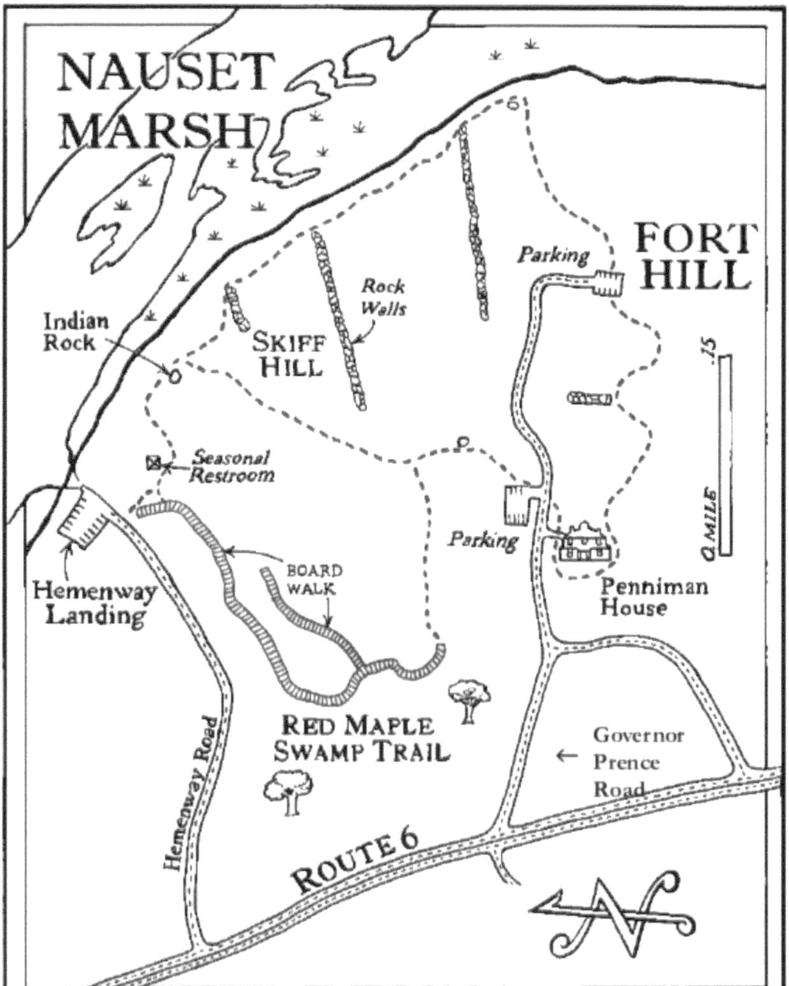

Figure 1.1. Fort Hill trail map, Cape Cod National Seashore. ***National Park Maps, http://npmaps.com***

The Wellfleet Conservation Trust provides maps, guided walks, and other methods to enhance access. All of the open space and conservation lands managed by the National Seashore, the town, and various conservation and environmental organizations are shown on a large, publicly accessible digital map.

At each trailhead of lands that the trust manages, a sign identifies it as the start of a trail within the conservation land. Each sign includes a QR (or quick response) code, for example, the one shown here. This is a machine-readable optical label for an internet link, or URL. It is essentially a two-dimensional barcode. With a smartphone camera, a user can find a trail guide with a map for the specific trail.

On a website and paper-based guides, one can find a map of each area with public walking trails marked. It includes descriptions of the land area, the trail distance, and geocoordinates (longitude and latitude). It has directions to get to the site from major roads or landmarks. It also includes a textual description, for example:

> Bayberry Hill Conservation Area and Trails: This relatively small parcel and short trail off of Lieutenant Island Road provides a surprisingly varied and tranquil landscape of field, forest, and marsh. . . . The culminating Salt Marsh view, in particular, stimulates the walker to look at the play of light and the great vistas, the flight and feeding of land and sea birds, and to savor the aroma of pine and sea.

This parcel description, brief though it is, reveals the possibilities for learning through a visit to the site. It includes suggestions regarding the varied ecosystems that come together in one area ("field, forest, and marsh"), the wildlife one might see ("land and sea birds"), and the senses that may be activated ("aroma of pine and sea"). The maps and guides serve instrumental purposes, for instance, showing the way, of course, but they also represent a portal to discovery.

Conservation as a School Activity

Wellfleet is a special case, with an unusually high percentage of protected land (70 percent), but the general idea of exploring one's neighborhood can be done anywhere. This can happen through informal learning, for instance, with family and friends, or in a library program. It can also be integrated with regular school activities.

For four years now, a group of volunteers and teachers have helped elementary-school children (grades 3–5) learn through the network of Wellfleet conservation areas and corresponding maps. The Town's Open Space Committee has built upon the trails to develop activity guides and organized field trips for elementary-school children.

Special purpose activities and field guides link learning in the school with learning in the outdoors. These materials and activities constitute an ecosystem for learning. Through a combination of classroom activities and field trips to actual trails, the children learn about the values and processes of conservation, compass directions, map reading, local flora and fauna (e.g., fiddler crabs), and access to and navigation of the trails.

They discuss reasons for exploring nature, including having fun, meditating, perhaps writing a poem, exercise, and learning about the town. One child suggested that a good use for the trails through the woods was for hiding! The network of map-related tools makes it possible for the children to extend the activities with their friends and families.

Children learn how to find the route, or multiple routes, from their school to an old wooden bridge or from their home to a trail. One discovered a trail very close to home that he had not been aware of before. They use the compass on their smartphones.

An especially popular activity within the overall project is to search for and identify objects, for example, scat, seashells, or particular types of trees. Detailed lesson plans are available, but in general the teachers take the lead because they need to integrate the trail activities with other parts of the school schedule.

Although all children appear to enjoy getting outdoors, experiencing nature, solving problems involving spatial reasoning, and exploring, some may benefit especially from these activities. A primary aim of the project is to connect with children who may feel left out of other activities.

The Conservation Trust activity guides were developed for a general audience—town residents, visitors, young and old people, as well as those with a variety of purposes. But the network creates a system of resources that can be employed in many ways, serving as a school activity, a way to involve parents in children's schooling, as well as a host of informal learning activities.

THE APPALACHIANS

A notable network of maps that supports informal learning is that for the Appalachian Trail. In 1921, Benton MacKaye recognized that the Appalachians were not just a natural marvel; they also shaped the economy and lifestyle of the region. He proposed the construction of a hiking trail as the connecting thread. This trail would follow the ridge crests, connecting farms, work camps, and study camps for urban residents. Now a continuously marked footpath leads from Springer Mountain, Georgia, to Mount Katahdin, Maine, a distance of more than 2,360 miles. Around three million visitors hike a portion each year.

Guidebooks, brochures, overall trail maps, maps for specific sections, topographic maps, digital maps, and more are available to guide the hiker. The connections among these maps, including common scale and naming conventions, make the network of maps far more useful than isolated maps would have been. Today, that network is being extended to the entire North Atlantic region.

A Brief History of the North Atlantic Region

The plate tectonics account of the Earth is full of massive shifts of land (entire continents and supercontinents), rotations, collisions and rifts, formations of oceans and mountain ranges. The term *continental drift* diminishes the violence and drama of these events, often invisible within a human lifetime but cataclysmic over geologic eras. Rather than taking the few million years of human evolution, they happen over many millions, even tens or hundreds of millions of years.

One need only consider the Ring of Fire, a region of the Pacific Ocean basin named for the many earthquakes and volcanic eruptions that occur there. It comprises a nearly continuous series of oceanic trenches and volcanic arcs and belts. The plate movements there account for this geologic activity.

One relatively recent part of this account is the story of Pangaea (figure 1.2). This was a supercontinent that existed during the late Paleozoic and early Mesozoic eras. It assembled from earlier continental units approximately 335 million years ago, and it began to break apart about 175 million years ago. Much of Pangaea was in the Southern Hemisphere (exactly the opposite of where most land mass is today) and surrounded by a super-ocean, Panthalassa.[10]

In 1912, Alfred Wegener observed from studying maps that the continents today fit together like a jigsaw puzzle. This led him to advocate for the theory of continental drift. Wegener was discredited by most geologists until his theory was revived in the 1950s due to paleomagnetic evidence.[11] The modern term for a theory with affinities to Wegener's continental drift is *plate tectonics*.

International Appalachian Trail

Today, the International Appalachian Trail (IAT) is a developing network of walking trails that extends through all of the geographic regions once con-

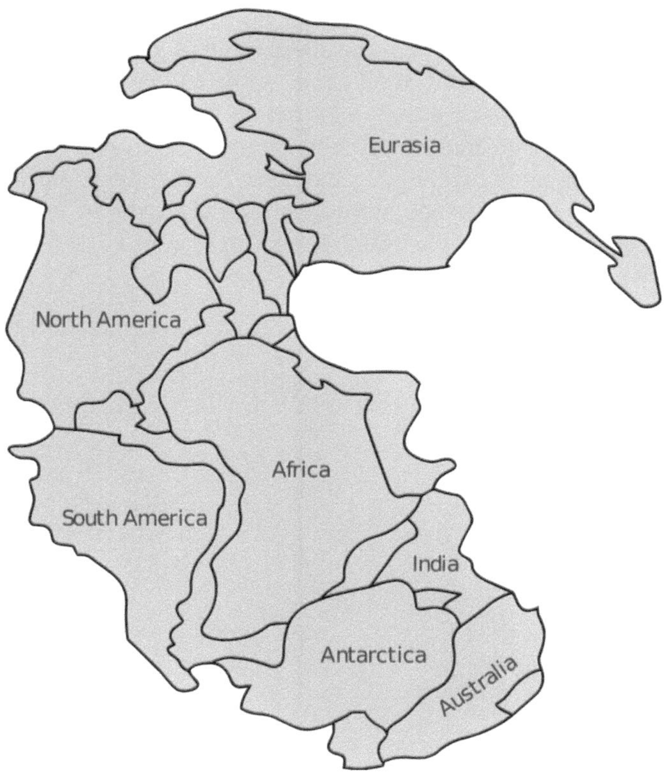

Figure 1.2. Pangaea supercontinent, circa 200 million years ago, before its breakup, with names of the modern continents. ***Wikimedia Commons, user: Kieff, 2009, CC BY-SA 3.0***

nected by the mountains of central Pangaea. On the western Atlantic side, it begins with the Florida Trail in Key West. That leads northward to Springer Mountain in Georgia, the southern terminus for the famous (U.S.) Appalachian Trail. It then follows the spine of the Appalachian Mountains to the northern terminus at Mount Katahdin in Maine. From there it winds through mountain scenery, pastoral countryside, and rugged coastline in maritime Canada to Crow Head on the north island of Twillingate in Newfoundland.

The trail continues to develop across the North Atlantic Ocean islands, through the Caledonides in Greenland, the Faroe Islands, Shetlands, Scotland, Ireland, and Wales, and on to Norway. It then heads south through western Europe into Morocco. An initial route there runs from near Marrakesh south to the ancient walled city of Taroudant. Eventually the trail could run a thousand miles or more along the crest of the Atlas Mountains, from the Mediterranean Sea to the Sahara Desert.

It will take many years to complete the trail and make it accessible to a large public. But the idea already inspires. Perhaps no one will ever walk every one of the thousands of miles on its spine and connecting trails. Yet, by linking the present with a time long before the first dinosaurs, crossing regions, countries, and continents, drawing upon multiple languages and cultures, venturing into ecozones of mountains, forest, marsh, meadow, farmland, coastal dunes, tropics and arctic, the IAT becomes an imaginary, yet tangible, path to experience the Earth we live on. Figure 1.3 shows the North American portion of the IAT.

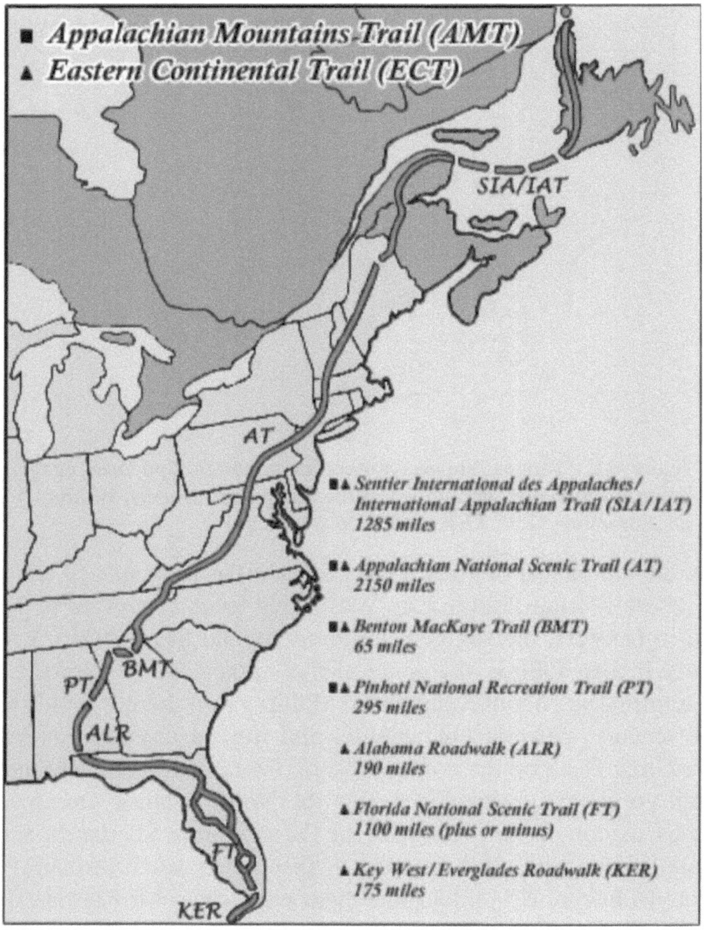

Figure 1.3. A portion of the International Appalachian Trail in North America. ***CC BY-SA 3.0***

The IAT organization presents a bold and comprehensive mission statement: "In addition to connecting people and places, the goal is to promote natural and cultural heritage, health and fitness, environmental stewardship, fellowship and understanding, cross-border cooperation, and rural economic development through eco and geo tourism."[12] Each of these goals represents opportunities to *learn*: about people, their homes, nature, cultural heritage, health, environmental consciousness, and more.

The Network of IAT Maps

We can imagine someday a massive trail guide for the entire trail from Key West to Taroudant. It would need to include not only the main spine of about fifteen thousand miles, but also an equal or greater amount of side trails and spurs. Also, a gigantic map could cover the entire system. It would need to be a large-scale map 1:25,000 (4 cm on a map = 1 km) or larger for the purpose of walkers. Even if such tools were to exist, they would be unwieldy.[13] Moreover, although the idea of the massive trail system is exciting, one might in practice prefer more localized guides reflecting the special characteristics of each ecosystem, region, or country.

Instead, we already have a network of maps, trail guides, photo galleries, through hike descriptions, and general information websites. Taken together, these resources make it possible for one to experience any portion of the vast IAT with relative safety and awareness of key features. The network makes it possible to customize the learning experience for children or adults, day hikers or through hikers, or those with special interests like birds, plants, cultural heritage, local history, geology, economic development, environmental conditions, or personal connections.

One could walk on the IAT without a map. Or one could use a single map and trail guide for a portion of it to gain a better understanding and appreciation of its physical and social history. With additional maps composing a network, one could see more clearly the connected story that it seeks to tell. Walking down a leaf-strewn trail with trees lining the way a walker could focus on the immediate forest sounds or smell; study wildflowers, fungi, or butterflies; meditate on life; or engage in a myriad of other purposes that visitors bring or discover along the way.

The network of trails and the associated maps not only provide useful information, but also generate all sorts of questions: If the rock formations in Newfoundland and the Scottish Highlands are similar, part of a single, ancient mountain range, is the flora similar as well? Can one (especially one who is not a professional geologist) makes sense of the rock formations and how they support the plate tectonics account? Is there any way to comprehend, as opposed to simply labeling with a number of years, the immensity of the time involved in forming this mountain range?

POLYNESIAN WAY FINDING

Shifting now from the Atlantic to the Pacific, and from walking to sailing, we can examine a very different kind of navigation, in this case, starting without maps as we know them. Centuries before European explorers began to explore the Pacific, the ancestors of today's Polynesians had sailed to every habitable island. They did this without maps printed on paper, which European sailors relied on, or their crucial instruments, for instance, sextant, compass, protractor, and hourglass.

Nainoa Thompson writes about how the early sailors managed to navigate successfully using nonconventional mapping.[14] The process begins with the star compass, the "basic mental construct for navigation." We have Hawaiian names for what are called the houses of the stars—the place where they come out of the ocean and go back into the ocean.

This navigation depends on knowing where you sailed from, including speed, direction, and time. This allows registration akin to pairing a known landmark with a map symbol in conventional navigation. Thompson describes the process as follows:

> When we go down to Tahiti, we have this mental image of our course line for the trip. We tend to try to follow it, and if we follow it properly, we will end up in what I call a "box." In this box, there are many islands. In the Tuamotu archipelago, we cannot sail into there and not find an island. This box is four hundred miles wide. The first part of the journey to Tahiti is not trying to get to Tahiti, but to make sure that you hit this box. And then we have to identify the island that we hit, and once we do that, we know the direction to Tahiti.

They also use average sail time based on wind speeds and conditions. For direction, they use the Sun when it is low on the horizon. We have names for how wide the Sun appears, and for the different colors of the Sun's path on the water. He continues,

> Sunrise is the most important part of the day. At sunrise you start to look at the shape of the ocean—the character of the sea. You memorize where the wind is coming from. The wind generates the waves. You analyze the character of the waves. When the sun gets too high, you steer by the waves. And then at sunset we repeat the pattern. The sun goes down; you look at the shape of the waves.
>
> Did the wind change? Did the swell pattern change? At night we use the stars. We use about 220 by name—where they come up, where they go down.

When the sky clouds over and way finders can use neither the sun nor the stars, they rely on the ocean waves. One experienced way finder describes how he can lie down inside the hull of the canoe and feel the different wave patterns. He can feel the canoe's direction with his body.

Finding low atolls is difficult, but wave patterns change when the canoe nears an island. The behavior of dolphins and other animals changes. Terns go out to sea at dawn and return in the evening to sleep. They go out about 130 miles. The flight path of the bird is the bearing of the island. One way finder noticed a small fish in a bird's beak. He knew that the birds were nesting, and they were taking food back before they fed themselves.

The diverse set of techniques that way finders employed look quite different from those of the European explorers, but they are equally complex and surprisingly effective, especially when compared to the European at sea without a printed nautical chart. The juxtaposition of Polynesian way finding with other techniques—using sextants and navigation charts, for example, or satellite-based global positioning systems—raises important questions.

It is tempting to say that way finders manage to navigate without maps. But might it be more accurate to say that they do use maps, which employ time, memory, and multiple media—stars, terns, wind, waves, and their own bodies as instruments? Finding our way always involves multiple sources of information and insight, including our memories, advice from others, and a sense of the territory.

What is the purpose of a map anyway? It helps us see where we are, but does it not also reflect our culture and values? Way finders carry on a tradition, which values understanding the ocean and the life around it, not simply finding the next island. Which values do we express when we use or create any map or means of navigation?

Moreover, maps are part of a political economy. Traditional way finding represents a form of resistance against otherwise overwhelming economic and cultural forces. For example, in the republic of the Marshall Islands, indigenous people have long struggled to maintain their way of life, even their very survival in the face of U.S. nuclear and missile testing and the military occupation of Kwajalein Island. Such ancient tools as stick chart maps are not simply instruments to solve technical problems, but important artifacts for self-determination.

HUMAN MIGRATION

At the L'Anse aux Meadows historic site in Newfoundland are many maps, as well as a museum exhibit, an outdoor exploration space with re-created

Norse buildings, and tours to help visitors understand the archaeological significance and the history of human migration.

Cousins

Early people came out of Africa into the Middle East. Some then headed east across Asia, Beringia, and North America, eventually making their way to Newfoundland and, ultimately, to L'Anse aux Meadows at its far northern tip. Others went north and west into Europe, eventually to Scandinavia. Some of those were Vikings or Norsemen, who ventured across the treacherous North Atlantic to Iceland, Greenland, and finally the northern tip of Newfoundland at L'Anse aux Meadows.

At that northern peninsula, the Vikings encountered the longtime Newfoundland residents who were their cousins from thousands of years earlier. The historic site helps re-create for the modern visitor the experience of two worlds meeting. Most of them join in collective astonishment at something they may have understood in principle but never fully imagined.

Standing at the site, the visitor may well be at the place where human migration out of Africa circumnavigated the globe for the first time. The experience reveals the importance of Norse travels, trade, and encampments as well as the journeys of the indigenous inhabitants of Newfoundland. Visitors learn of the importance of the human encounter and how it became for the Norse the gateway to Vinland. They also learn how archaeologists came to understand the stay of the Norse and their ultimate abandonment of the encampment.

This understanding is made possible by the use of maps showing the migrations out of Africa; the movements across Asia, North America, and Europe; and, eventually, the settlements of the two worlds in L'Anse aux Meadows. Once at the site, a visitor can find maps of the visitor center, the Norse settlement, and individual Norse buildings.

The site at L'Anse aux Meadows was found again in 1960. More than fifty years later, a similar Viking site was discovered at Point Rosee (formerly known as Stormy Point) in the southwest of Newfoundland, more than four hundred miles away. "Space archaeologist" Sarah Parcak used satellite maps to find evidence of human activity in the landscape.

Satellite images showed a faint difference in the vegetation in the form of a rectangle. On-site investigations revealed turf walls and a hearth. The site has bog iron, natural deposits of the metal that would have been very attractive to Vikings. The iron had formed as rivers carried ore from the mountains into wetlands, where bacteria leached iron from water and left behind the metal.

This is the southernmost and westernmost location in the Americas where evidence of pre-Columbian ironworking has been discovered.

Instead of mining, the Vikings usually harvested iron from peat bogs. One reconstructed Norse longship needed seven thousand nails. To produce the necessary 880 pounds of iron would require smelting thirty tons of raw iron ore. Research is still underway to confirm the Point Rosee site, but archaeologists say that the structure there differs from both Native construction and that of Basque fishermen who visited in the sixteenth century. The only people who would have been working with bog iron were the Vikings.

Satellite maps showing human-made structures, maps of river flow delivering iron, maps of ocean currents and travel routes, maps of structures and settlements, all play a role in understanding these sites and the early encounters between the long-lost cousins of Europe and North America. They also demonstrate the enormous potential of using maps for thinking and learning.

Recent Encounters

A similar encounter occurred in Papua New Guinea much more recently. People from Southeast Asia moved into the highlands there fifty to sixty thousand years ago during an Ice Age period when the sea was lower and distances between islands was shorter. Then, in 1933, gold prospectors came from Australia, in particular Michael Leahy and his brothers. They led a series of prospecting expeditions into the highlands and initiated the first contacts between highlanders and Europeans.

Leahy's ancestors were originally from Europe, the Middle East, and Africa. Again, long-lost cousins had an encounter. This one is more fully documented with field notes, photographs, film, and text.

A large gulf existed between the two cultures, and large misunderstandings on both sides. Kirupano Eza'e, one of the highlanders, said of the first encounter,

> Once they have gone . . . the people sat down and developed stories. They knew nothing of white-skinned men. We had not seen far places. We knew only this side of the mountains. And we thought we were the only living people. We believed that when a person died, his skin changed to white and he went over the boundary to "that place"—the place of the dead. So when the strangers came we said: "Oh, these men do not belong to the earth. Let's not kill them—they are our own relatives. Those who have died before have turned white and come back."[15]

Eza'e shows through his account how exploration can both enlighten and obscure in one encounter. He acknowledges the great discovery that the highlanders were not "the only living people." Moreover, although the strangers were indeed strange, he recognizes them as "our own relatives." Nevertheless, he and his compatriots could not accept the strangers as simply human; they were, instead, those who have died before, turned white, and come back.

Mark Twain famously writes, "Travel is fatal to prejudice, bigotry, and narrow-mindedness, and many of our people need it sorely on these accounts. Broad, wholesome, charitable views of men and things cannot be acquired by vegetating in one little corner of the earth all one's lifetime."[16] His writings reveal some appreciation of the *other*, at least for his time.

However, they also show a disconnect; travel alone cannot banish "prejudice, bigotry, and narrow-mindedness." As one example of many, when Twain speaks of the Azores, he writes, "The community is eminently Portuguese—that is to say, it is slow, poor, shiftless, sleepy, and lazy."[17] Any broader view needs reflection and analysis to develop beyond the encounters of travel alone.

Political Struggle

Being mapped means being recognized, by the cartographer first, and then by current and future users. That recognition always has consequences, sometimes minor and sometimes enormous. Denis Wood says, "Power is a measure of work. Which is what maps *do*: they *work*."[18] He elaborates that they work both in the sense of effectively solving a practical problem and in the sense of applying social forces, authority, to people.

If the real estate agent does not include a nearby garbage dump on a map, that may lead a house buyer to purchase a home he would not have chosen otherwise. But if the heritage land of a community is left out of an official map, one that represents only the perspective of the dominant culture, the omission could destroy the culture of that community, their livelihood, and even their lives.

When European explorers led the way to global colonialism, they conveniently left many areas of their maps blank, suggesting that the only question was which European nation should conquer it:

> More ominously, Native American tribal areas were not included on early European maps of the Americas, giving readers of those maps the impression no one lived there—at least, no one of consequence. No landowners. These are the kinds of blanks that fire Marlow's imagination in Conrad's *Heart of Darkness*—the blanks that certain minds found to be a call for colonialism and conquest.[19]

In *Imagined Communities*· Benedict Anderson discusses how we create communities and nations through our discourse and our technologies. He shows how maps, museums, and the census shaped the grammar that made possible the creation of nation states and peoples, for example, “Burma” and “Burmese.” “This style of imagining did not come out of thin air. It was the product of the technologies at navigation, astronomy, neurology, serving, photography and print, to say nothing of the deep driving power of capitalism.”[20] The imagining furthered the project of colonialism.

Today, indigenous peoples around the globe, including Native Americans, Australian Aboriginals, and tribes of Indonesia and Tanzania, face the converse challenge: they simply want the state to see them. Although these people may have been residents for centuries, they may not retain modern legal ownership. Bernard Nietschmann writes, “More indigenous territory has been claimed by maps than by guns. This assertion has its corollary: more indigenous territory can be reclaimed and defended by maps than by guns.”[21]

Richard Chase Smith and his team with Instituto del Bien Común in Lima have worked with local groups over twenty years to map twenty-five hundred native communities in Peru’s Amazonian region. In 2005, they also began to map the country’s eight thousand *comunidades indígenas*, or peasant communities. This effort meant that communities could, for the first time, declare their fundamental existence to the Western establishment and the power it holds.

This type of work has grown, with the launch in 2015 of LandMark.[22] The platform seeks to help indigenous communities protect and secure their land rights by making their lands and the peoples who live on them visible. The indigenous and community land maps represented there cover 12.4 percent of the world’s land, out of an estimated 50 percent or more held by indigenous peoples and communities. The site includes data on land tenure, potential pressures on lands, changes in land cover, and contributions to protecting the environment. Various analysis tools can be used to calculate forest cover loss, carbon stocks, concessions, and the extent of other assets and threats to the land.

The centerpiece is a zoomable, interactive world map. Users include communities, governments, companies, development agencies, and researchers. Map regions are color coded: *Indigenous land* is collectively held and governed by indigenous peoples with distinct social, cultural, or economic characteristics; *community land* is collectively held by a community (a grouping of individuals and families that share common interests in a definable local land area) regardless of recognition under statutory law.[23]

THE LORE OF PLACE

In a chronicle of the Bank Street College of Education approach to field trips, Salvatore Vascellaro argues that education should be more than an intellectual pursuit; it should move the learner outward physically and socially.

The college is well known for its intensive use of maps to aid exploration. Describing how this works for first- and second-grade children in New York, he writes, "We began with walks in our Manhattan neighborhood: on the school block, around the block, to the avenue filled with small stores, over to the East River, into Central Park."[24] His account emphasizes that getting out of the classroom unlocks the potential for both students and teachers.

Maps are both initial guides for venturing out and tools for making sense of experiences. They cannot capture the full multisensory experience—the warmth of an early spring sun, a breeze coming off the pond, the smell of flowers or a factory, the sound of an owl—but they can be an entrée into discovery and a means for making sense of it.

Expanding on this idea, Lucy Sprague Mitchell established the annual Long Trip in 1935. The first trip, to Morgantown, West Virginia, was designed to expose student teachers to new physical, social, and political environments and expand their concept of human geography. Students learned firsthand about caves filled with coal, mountains of slag, contaminated rivers, the dependence of the economy on production of electricity from coal, strip mining, unions, and much more. As Mitchell said, "There is also the learning that comes from a living gripping experience where a feeling tone is added to thought."[25]

In recent years, Long Trips have gone to New Orleans to study environmental injustice, including the devastating impact of Hurricane Katrina; to Patzcuaro, Mexico, to learn about the arts, culture, and ceremonies of indigenous peoples; to the Scotland Forest Schools where young children spend most of the school day in the forest.[26] Throughout, the emphasis has been on getting to know a place, using and making maps, but especially just being open to "a living gripping experience where a feeling tone is added to thought."

This is similar to what Eamonn Wall means with the Irish term *dinnseanchas*, or "lore of place."[27] The individual is formed by interaction with place as story. That story reveals how we are a part of place, engage with it, and learn from it.

Dinnseanchas recalls a passage from Oliver Wendell Holmes about viewing his own life as a "dissected map," much like a jigsaw puzzle:

> The life of an individual is in many respects like a child's dissected map. If I could live a hundred years, keeping my intelligence to the last, I feel as if I could put the pieces together until they made a properly connected whole. As it is, I, like others, find a certain number of connected fragments, and a large number of disjointed pieces, which I might in time place in their connection. . . . What strikes me very forcibly is the arbitrary and as it were accidental way in which the lines of junction appear to run irregularly among the fragments. . . . Blanks which have been left in former years find their complement among the undis-

> turbed fragments. If I could look back on the whole, as we look at the child's map when it is put together, I feel that I should have my entire life intelligently laid out before me.[28]

Map–journey–story–life: the lore of place helps us make sense of a particular locale. Our travels enable us to restore the dissected map into a unified whole. That map, in turn, helps us understand our role in that place. Yes, maps are useful tools to help us locate ourselves and to traverse the place but, more significantly, they are the means we employ to make sense of our life in context.

NOTES

1. David Farrier, *Footprints: In Search of Future Fossils* (New York: Farrar, Straus & Giroux, 2020), 4.
2. Farrier, *Footprints*, 28–29.
3. Rebecca Solnit, *Wanderlust: A History of Walking* (New York: Verso, 2002).
4. "Miles of Public Roads: Los Angeles County," *Los Angeles Almanac* (blog), n.d., https://www.laalmanac.com/transport/tr01.php.
5. David Millward, "Sale of Road Atlases Soars as Satnav Drivers Seek Extra Help," *The Telegraph*, September 6, 2013, https://www.telegraph.co.uk/motoring/news/10292351/Sale-of-road-atlases-soars-as-satnav-drivers-seek-extra-help.html; Robert Duffer, "Why the Road Atlas Still Matters," *Chicago Tribune*, August 6, 2016, https://www.chicagotribune.com/autos/sc-road-atlas-autotips-0804-20160803-story.html.
6. Even with zooming, a screen does not show well the broad layout of roads; an atlas can give a sense of scale, scope, and connectedness.
7. Humble Oil put out road maps for many years. The company was founded in 1911 in Humble, Texas. It operated there and in neighboring states for many years. Through mergers and rebranding, it became part of Exxon in 1973. In 1957, scientists from Humble Oil published one of the first oil industry studies tracking "the enormous quantity of carbon dioxide" contributed to the atmosphere since the Industrial Revolution "from the combustion of fossil fuels." One can learn from the publication, as well as the map itself.
8. Edward B. Espenshade and Joel L. Morrison, eds., *Goode's World Atlas* (Chicago: Rand McNally, 1974), vii.
9. These include the Cape Cod National Seashore, Mass Audubon's Wellfleet Bay Wildlife Sanctuary, the Wellfleet Conservation Trust, and the Town of Wellfleet's Open Space.
10. Pangaea formed through complex collisions and separations of continental masses, with the subsequent alteration of mountain ranges and oceans. Evidence shows that the Pangea integration contributed to environmental deterioration exacerbating the greatest catastrophe for life on Earth, the Permian extinction (about 250 million years ago).

Much later, the continents of North America and Eurasia split from Pangaea, opening up the Atlantic Ocean between them. Rocks in the Appalachian Mountains of North America, the Little Atlas of Morocco, and much of the Scottish Highlands are similar and provide evidence for Pangaea and its major mountain range. These are among the oldest mountain systems on Earth.

11. Michael McIntyre, H. Peter Eilers, and John Mairs, *Physical Geography* (New York: Wiley, 1991), 273; Ronald N. Giere, *Explaining Science: A Cognitive Approach* (Chicago: University of Chicago Press, 1988).

12. The International Appalachian Trail website, https://www.iat-sia.org.

13. A region including the entire trail would be more than seven thousand kilometers from Key West to the Sahara Desert and four thousand kilometers from south to north. A paper map at the 1:25,000 scale would cover about six football fields and not fit in anyone's backpack.

14. Nainoa Thompson, "Non-Instrument Navigation," *Wayfinders: A Pacific Odyssey* (blog), 1999, https://www.pbs.org/wayfinders/way finding2.html.

15. Bob Connolly and Robin Anderson, *First Contact: New Guinea's Highlanders Encounter the Outside World* (New York: Viking Penguin, 1987), 6.

16. Mark Twain, *The Innocents Abroad or The New Pilgrim's Progress* (New York: New American Library, 1966), 512.

17. Twain, *Innocents Abroad*, 33.

18. Denis Wood, *Rethinking the Power of Maps* (New York: Guilford, 2010), 1.

19. Peter Turchi, *Maps of the Imagination: The Writer as Cartographer* (San Antonio, TX: Trinity University Press, 2004).

20. Benedict R. Anderson, *Imagined Communities: Reflections on the Origin and Spread of Nationalism (revised and extended)* (London: Verso, 1991), 185.

21. Bernard Nietschmann, "Defending the Miskito Reefs with Maps and GPS," *Cultural Survival Quarterly* (1995): para. 20.

22. "Our Impact," *Rights + Resources Initiative Coalition* (blog), n.d., https://rightsandresources.org/en/impact-stories/.

23. Karen Emslie, "This Alternative to Google Maps Aims to Protect Indigenous Land," *Medium One Zero*, June 14, 2019, https://onezero.medium.com/this-alternative-to-google-maps-aims-to-protect-indigenous-land-62e2e414eb04.

24. Salvatore Vascellaro, *Out of the Classroom and into the World: Learning from Field Trips, Educating from Experience, and Unlocking the Potential of Our Students and Teachers* (New York: Bank Street College of Education, 2011), xv.

25. Vascellaro, *Out of the Classroom*, vii.

26. https://graduate.bankstreet.edu/cps/travel-programs/long-trip/history/. One could applaud Bank Street for this program while at the same time noting that most students in the world, or even the United States, have no chance to participate in such a program. Nevertheless, it stands as a model for the value of connecting classroom learning to the larger world and could inspire alternatives that might be achievable in a wider range of settings.

27. "Deepmaps: Reading Tim Robinson's Maps of Aran," *Terrain Org: A Journal of the Built and Natural Environments* (Spring/Summer 2012).

28. John T. Morse Jr., ed., *Life and Letters of Oliver Wendell Holmes* (London: Sampson Low, Marston & Co., 1896), 28.

2

Constructing

"What a useful thing a pocket-map is!" I remarked.

"That's another thing we've learned from your Nation," said Mein Herr, "map-making. But we've carried it much further than you. What do *you* consider the *largest* map that would be really useful?"

"About six inches to the mile."

"Only *six inches*!" exclaimed Mein Herr. "We very soon got to six yards to the mile. Then we tried a *hundred* yards to the mile. And then came the grandest idea of all! We actually made a map of the country, on the scale of a *mile to the mile*!"

"Have you used it much?" I enquired.

"It has never been spread out, yet," said Mein Herr. "The farmers objected: They said it would cover the whole country, and shut out the sunlight! So we now use the country itself, as its own map, and I assure you it does nearly as well."

—Lewis Carroll, *Sylvie and Bruno Concluded*[1]

Lewis Carroll's final novel is a fantasy with Carroll's characteristic insights about maps. After imagining a country map the size of the country itself, Sylvie and Bruno go on to consider the smallest world they would care to inhabit, perhaps "a little teeny-tiny world," just large enough for the two of them. Mein Herr points out, "Then you would have to stand on opposite sides of it . . . and so you would never see your sister *at all*!" Fortunately, no one has ever made a map large enough to cover an entire country and shut out the sunlight,[2] or a world so small that we could never see our neighbors.

But Carroll's fantasy finds parallels in the everyday world. A school playground has a map of the United States large enough for children to inhabit different states and get a semblance of what it means to walk across the country.

A nature trail for visually impaired people has a rope to mark itself and braille or audio text describing each portion as the visitor encounters it.

Could an entire city become a map for a country? What would it be like for the city not only to go about its ordinary business, but also tell a story about itself and its history? Considering these questions gives us insights into how maps are a means for *constructing* the spaces we inhabit, not simply describing them.

HERMENEUTICS OF MAP CREATION AND USE

Maps are useful tools for finding our way or knowing where we are in relation to the people, objects, and places important to us. Any map, to be useful, needs to be read in the sense of ground-level interpretation of its symbols, orientation, and scale.

A focus on the technical function of a map may obscure that it is also a rhetorical device conveying a message, with different meanings for the mapmaker and different map users. It is interpreted in light of the user's experiences and purpose and, perhaps, the user's inferences about the mapmaker's intentions. This is all to say that effective use of a map is a *hermeneutic process*, in which meaning is created, shared, ignored, or transformed.[3]

Think of the example of Ukraine today. How would its map be drawn by a cartographer in the United States, the European Union, Ukraine, or Russia? Would Crimea be colored as part of Ukraine? Of Russia? How about Tuzla Island or far eastern Ukraine? The cartographer would make decisions based on his understanding of the situation, not only on the intended effect on a user. Following that, the user would interpret the coloring, with nonchalance, anger, puzzlement, pleasure, or other emotions based on his experience and purpose.

The Ukraine map (figure 2.1) from the U.S. Central Intelligence Agency's *World Factbook* shows the full outline of Ukraine with disputed areas colored in. One can track changes in Ukraine in recent years through a series of maps from the *World Factbook's* archives and now in the histories through such online sources as Wikipedia. The coloring of disputed areas shifts both as conditions on the ground change and as cartographers and their clients seek to present the world in a particular way. The coloring is accurate for some purposes but probably pleases no one. Nevertheless, the series of maps of Ukraine becomes a tool for learning about recent history in the area, as well as a test of one's understanding of that history.

Figure 2.1. Disputed regions of Ukraine. ***U.S. Central Intelligence Agency, World Factbook, https://www.cia.gov/***

Hermeneutics is a broad and complex field. For our purposes, a key principle is that understanding is based on *interpretation*. Through the process of engaging with and using a map, the user creates meaning that is a *fusion of the horizon* between her world and that of the mapmaker.[4]

STREET NAMES IN PARIS

The street signs of Paris are a unique and attractive feature that any visitor notices. Street names are indicated on every angle of every building on every corner in stark white letters on blue enameled metal placards, set against medieval brickwork. Stone tablets had been replaced by iron signs with white letters on a black background in 1823. The deep blue background came twenty years later. Enameled porcelain was introduced just a few years later. Most of the signs indicate the arrondisement.

For example (figure 2.2), in the Marais one can find Place des Vosges, the oldest square in Paris. It is named for the Vosges, a range of low mountains in eastern France, near the border with Germany. An early example of city planning, it is a perfect square (140 meters) and is the prototype for subsequent residential squares in Europe. Madame de Sevigné, Cardinal Richelieu, Victor Hugo, Théophile Gautier, Alphonse Daudet, Georges Simenon, and other notables lived here. During the French Revolution it was called Place des Fédérés (Federated Square) and Place de l'Indivisibilité (Indivisibility Square).

Figure 2.2. Street sign for Place des Vosges in the 4th arrondissement. ***Wikimedia Commons, Marko Kudjerski from Toronto, Canada, CC BY 2.0***

Street names (formally called *odonyms*, from the Greek *hodós*, or "road") were first used in France in the late thirteenth century. The early names were often long and unwieldy descriptions. For example, in Lyon in 1342, one road was called la rua qui fiert vers los Fraros Prejurs (the street that comes out near the preaching friars). The names were a way of providing a geographic description where a map was lacking, perhaps not so different from the GPS narrative one hears with smartphone directions.

Throughout time, these lengthy descriptions were gradually shortened, especially as street names and signs were adopted more fully. But it wasn't until the eighteenth century that signs were actually mounted on most street corners. Before then, people found their way using hotel and shop signs, or local informants.[5] People in many countries today still navigate in that way, because no mounted signs are present and even online tools prove inadequate.

Initial street names were adopted by local consensus, but in the seventeenth century municipal authorities began to invoke famous French characters from politics, royalty, and the arts. After the 1789 Revolution, such street names as rue de la Liberté, Place de l'Égalité, and Route de la Nation sprang up throughout France. Thousands of royal and religious street names were removed in favor of revolutionary heroes and admonitions. The abbé Gregoire called for it to be "tout republicaniser": Notre-Dame would be the Place de l'Humanité républicaine, framed by the rue de la Sensibilité and rue de Génerosité. The new names were to be short, comprehensible, euphonic, and above all, morally correct.[6]

When Napoleon came to power in 1799, he ensured that he would be remembered with rue de Rivoli and Place d'Austerlitz (great victories), rue Massena (his favorite general), and such. He recognized the political power of these mini-historical monuments. Contention over names continues. Place Général de Gaulle in Marseille has had nine different names since 1778 according to the whims of those in power.[7] Throughout time, famous generals like Avenue Hoche (Lazare Louis Hoche, 1768–1797) and such battles as Avenue de Wagram (French victory over Austria on July 6, 1809) began to radiate from the Arc de Triomphe.

In 1911, the municipal council of Paris passed a bill requiring that the historical significance and dates of events should be added to street name signs. They wanted the streets of the city to "teach history to Paris children." Instead of a future sign reading simply rue de Rivoli, it would say rue de Rivoli/victoire française en 1797 (only on selected corners). Similarly, signs would read as follows:

- Avenue Victor Hugo, poète et romancier français, 1802–1885 (French poet and novelist)
- Rue Lincoln, célèbre président des États-Unis, 1809–1865 (famous U.S. president)
- Rue Descartes/1596–1650/Philosophe, Mathematicien et Physicien (philosopher, mathematician, and physicist)
- Rue du Chevalier de Saint-George/ancienne rue Richepance/1739–1799/ Musicien et chef d'orchestre/Colonel de la Garde Nationale (formerly named for Antoine Richepanse, now honoring Chevalier de Saint-George, musician and orchestra leader)

One can note that the last example records the street name's own history. The street was once named for Antoine Richepanse. He was a French revolutionary general, who as colonial administrator restored slavery to Guadeloupe after it had been abolished by the revolution in 1794. This kind of change reveals changing values. Presumably, more recent odonymists were less inclined to honor a revolutionary hero right or wrong, especially when his later work was to expand slavery.

Paul Escadies, a deputy, saw the potential for education in the street sign inscriptions:

> The picturesque and varied streets of Paris furnish a good education. The street names are mementos of our history—literary, scientific, artistic, political, and military. The committee on Paris inscriptions have placed marble tablets on the houses where illustrious men and women have been born and where they have lived and died.[8]

Living inside a Map

The street name signs of Paris are a marvel equal to any of the monuments there. They are highly functional, helping the millions of visitors each year to find their way. Few of those visitors fail to note that they are beautiful and distinctive, with attractive lettering on blue enamel plates. But in addition, the signs constitute a map that one can enter to learn about the city, its history, its language, and the scientific, literary, and artistic achievements of Western civilization.[9] One might argue that any city's street name signs do this, but Paris stands out, as it does in many other areas.

It is, thus, no surprise that Paris invites the *flâneur*, Charles Baudelaire's artist-poet of the modern city:

> The crowd is his element, as the air is that of birds and water of fishes. His passion and his profession are to become one flesh with the crowd. For the perfect *flâneur*, for the passionate spectator, it is an immense joy to set up house in the heart of the multitude, amid the ebb and flow of movement, in the midst of the fugitive and the infinite.[10]

For the *flâneur*, a wanderer, the rewards are many.

A daguerreotype (figure 2.3) that Louis Daguerre made in 1838 shows the Boulevard du Temple. The boulevard crosses rue du Temple, one of the oldest streets in Paris, linking rue de Rivoli and Place de la République. These detailed photographs map Paris as well as a standard cartographic map. With additional images Daguerre could have produced a complete image map of Paris, helping the visitor navigate to ancient sites in the city.

Temple refers to the commandery of the Knights Templar. They had set up their headquarters in Paris in 1140, building a fortified city in what is today the Marais. Remnants of fortifications dating back to the time of King Philip II (king from 1180 to 1223) can still be found on rue du Temple.

This photograph is the earliest surviving photograph showing a living person. The street is busy, but because the exposure lasted for several minutes the moving pedestrian and vehicle traffic left no trace. Only the two men near the bottom left corner, one apparently having his boots polished by the other, stayed in one place long enough to be visible.

Figure 2.3. Daguerreotype made by Louis Daguerre in 1838, showing the Boulevard du Temple. The Photography Book *(London: Phaidon, 1997). Public domain*

In addition to showing living people, the daguerreotype hints at modern aerial imaging. With enough plates and time, Daguerre could have shown many streets, their orientation, width, junctions with other streets, and major buildings. Like contemporary aerial or satellite imagery, it is both an image of the world and a map, albeit without a legend.

Characteristics of Paris Street Names

Paris street names and accompanying descriptions provide a rich map for learning. Anywhere else Réamur street would be just a name, informing you only of the street you were on. But walking on rue Réamur in the 3rd arrondissement of Paris, you could learn from the very street signs that René Antoine Ferchault de Réaumur (1683–1757) proved that corals were animals, not plants. That might enlighten you about Réaumur the scientist and reveal a curious fact about nature as well.

You might be inspired to learn of his many achievements: Réaumur proved that the strength of a rope is more than the sum of the strengths of its separate strands; reported on gold-bearing rivers, turquoise mines, forests, fossil beds, and bird's nests; showed that the amount of carbon is greatest in cast iron, less in steel, and least in wrought iron; developed a new thermometer and temperature scale; showed that echinoderms could replace their lost limbs; was a founder of ethology; wrote on the possibility of spiders being used to produce silk; observed wasps making paper, which led to changes in paper-making techniques; and much more. You would be unlikely to follow that trail if you had not made friends with a street whose signs talked to you.

Walking along rue Réamur, you might notice the National Conservatory of Arts and Crafts, a doctoral degree-granting engineering school dedicated to education and research for science and industry. Going further, you might discover that many other scientists and engineers find their street near the Conservatory.

Rue Montgolfier honors the brothers Joseph-Michel (1740–1810) and Jacques-Étienne Montgolfier (1745–1799), inventors and pilots of the first hot-air balloons. Again, you might not have known about their work, or much about hot-air balloons, but simply walking the streets invites you to learn more about them, which you can do in the nearby Conservatoire.

Many streets named after doctors and medical researchers, for example, rue Jenner, can be found in a medical area in the 13th arrondissement. Rue Aristide Briand is in the 7th, running by the National Assembly, the European Commission, and other government offices, and streets named after other statesmen. This is a congenial spot for a man who served eleven terms as prime minister during the Third Republic, promoted the Kellogg–Briand Pact of 1929, helped establish a European Union, and won the Nobel Peace Prize. His description on a small blue sign opens the door to learning about him, government functions, and significant twentieth-century history.

Rue La Pérouse in the 16th (Jean-François de Galaup de La Pérouse, 1741–1788) honors the famous French navigator. Anywhere La Pérouse traveled, he found people who could draw a map in the sand. But his maps were "immutable mobiles." They were permanent (transferred to paper) and mobile (portable in a map case to take back to the court at Versailles). Bruno Latour argues that this technology of mapping provided an understanding of remote lands and consequent power over them. It enabled and furthered the development of the Empire.[11]

The municipal database describes the area around rue La Pérouse as hosting names of navigators. One example is rue Dumont d'Urville (Jules Sébastien César Dumont d'Urville, 1790–1842, who explored Australia, New Zealand, and Antarctica). Perhaps it also includes the nearby streets named for astronomers, rue Galilée and rue Kepler, or should we say that solar system exploration is a different cluster?

Is it significant, or only coincidental, that rue Galilée crosses Avenue d'Iéna? The latter is named after Napoleon's victory at Jena (Germany) in 1806, but Jena is also a leading city for scientific research. Optical instrument makers there—Carl Zeiss, Schott, and Jenoptik—continue the tradition of Galileo with telescopes and planetariums. The interpreter of the street names may see things that the urban designers did not.

Similar clusters exist for geologists, painters, musicians, novelists, and others. Nearly a hundred streets or squares are named after mathematicians alone, including many who are not French. Rue Newton, rue Galilée, and rue Euler come together at Avenue Marceau near the Arc de Triomphe.[12]

Becoming aware of what the street names represent and how the naming is tied to locations with meaning leads one to look for additional connections, which may be deliberate or serendipitous. Once the idea of clustering enters the head, it is impossible not to look for it everywhere.

Moreover, the names link to physical location. Consider the case of General Leclerc de Hauteclocque. He was a leader in the Free French cause and led the liberation of Paris and Strasbourg. On August 25, 1944, at the Montparnasse train station, he accepted the surrender of the German troops. Leclerc was also a staunch anti-Communist, opposing, among others, Jean Pierre Moulin, an alleged Communist. Moulin was head of the National Council of Resistance and was arrested, tortured, and killed in 1943. Each man has a major avenue in his honor: Avenue du Général Leclerc and Avenue Jean Moulin.

These avenues might have been in other parts of the city; they could have been parallel, or short and non-crossing. But instead, they meet in the 14th. The intersection is in the south of Paris, on the path the armies of liberation would have taken in 1944. Coming upon that crossing, one feels that Leclerc and Moulin, opponents earlier, finally came together in common cause, just as their streets do today.

The streets, their position relative to one another, and their place in the context of the geography of Paris, tell a story inviting further inquiry. They do not simply identify a line on a map or a postal address.

It is no coincidence that the Paris Opéra is on Boulevard des Italiens in the 9th. Nearby one can walk on rue Rossini (Gioachino Antonio Rossini, 1792–1868, famous for his thirty-nine operas), rue Gluck (Christoph Willibald Gluck, 1714–1787, composer of Italian and French opera in the early classical period), rue Auber (Daniel François Esprit Auber, 1782–1871, one of the founders of the grand opera genre), rue Halévy (Jacques-François-Fromental-Élie Halévy, 1799–1862, known for his opera *La Juive*), and rue Scribe (Augustin Eugène Scribe, 1791–1861, a French dramatist and librettist of many of the most successful grand operas). Note that it is not simply musicians per se, or composers, who reside near the Opéra, but composers of operas.

In another arrondissement, the 13th, the short (1/10 mile) rue Albert Einstein (identified as 1879–1955/Physicien Amèricain/D'origine Allemande/ Prix Nobel 1921, American physicist, born in Germany, Nobel Prize) is home to several important structures. Paris Diderot University has three of its newest buildings there. The Olympe de Gouges building houses facilities for the engineering school. It is named for one of France's first feminist writers, who wrote an antislavery play, a treatise on the rights of women, and argued for rights for so-called illegitimate children. A building named for Sophie Germain (French mathematician, physicist, and philosopher who developed elasticity theory) is home to mathematics and computer facilities.

On the same street, the unusual Tower of Biodiversity rises fifty meters high in a city that normally limits building height to thirty-seven meters. Edouard François designed it to show that high-rise structures can be used to create environmentally friendly cities with ample plant and animal life.

The street both honors and redeems Einstein. He had revolutionized our understanding of space and time, gravity, light, and other topics. He was also a fan of Paris and lectured at the University of Paris in 1922, so it was entirely fitting that he have a street named after him. But Einstein is also associated in many people's minds with $E = mc^2$, a formula that underlies the atom bomb and an era in which few women were allowed to pursue careers in science, especially in physics.

It seems that Paris and Paris Diderot University chose to feature the genius of Einstein, his forward thinking, and his basic humanity. The implication of their design is that a twenty-first-century Einstein would applaud the inclusion of women in math and science and the orientation of science to restoring the climate and biodiversity.

Some streets, for example, Avenue Victor Hugo, reflect the importance of the namesake. The avenue (more than a mile long and 118 feet wide) cuts across neighborhoods. The writer attained such a significance in French literary and political history that he requires a major street with Place Victor Hugo being a hub for ten streets.

It is clear that the names are intended to indoctrinate. Once the purpose was to teach about Roman Catholic religion and the royalty of France. Later, it was a means to celebrate the Revolution and promote its ideals. Napoleon transformed it further to honor himself and the scientists whom he revered. Later municipal and national governments used it to further other always shifting ends. For example, Square Louise-Michel in the 18th honors Louise Michel, the brilliant educator and leader of the Paris Commune of 1871, who had once been branded a traitor.

The effectiveness of the street naming is amplified by the fact that they are part of a network. Descriptions on the signs are supplemented by historical markers, most recently in the shape of a shield on a post. One can walk on rue Cyrano De Bergerac in the 18th and also read his historical marker outside the College of Beauvais. Savinien Cyrano de Bergerac (1619–1655) was a musketeer, duelist, and writer. He was also a disciple of the astronomer Gassendi. His plays, including satires of religion and absolutism, prefigured the Enlightenment. A fictionalized version of Bergerac was celebrated in a play by Edmond Rostand.

Today, one can look up rue Cyrano De Bergerac in online maps. The historical markers are indexed at waymarking.com, with GPS coordinates, images, descriptions, and metadata. The online resources provide seemingly unbounded information, obviously extending what can be done with a short text. However, walking on the street commemorating some person or event, or reading a physical marker situated in a significant location, cannot be matched in digital form, even with virtual reality. Standing outside where Bergerac's college stood on rue Jean de Beauvais or exploring his neighborhood conveys a sense of place. This is especially the case in a city like Paris, which has preserved much of its historical architecture.

As mentioned earlier, one can easily move from a scientist's or artist's street name to a museum providing more information about him or her. The brief descriptions on the occasional street sign allow one to bridge easily to a Web search. The street names become the center of an ecosystem of learning, connecting the pedestrian (do not try this while driving) with other streets, museums, musical venues, government buildings, monuments, books, websites, and more.

Streets without Names

The didactic intentions of the naming of streets in Paris are evident in ongoing controversies. At least one place is *sans nom* (nameless). The debate about whether and what to name it reflects the competing purposes of cartographers, residents, business owners, and others.

In the 12th arrondissement is a *carrefour* (crossroad) at the intersection of Boulevard de Reuilly (route to the old castle of Reuilly, residence of the Merovingian kings) and Boulevard de Picpus,[13] rue de Picpus, and rue de Louis-Braille (creator of the alphabet for the blind). The residents in the area wanted a pleasant space with fewer cars and more pedestrians. They sought more activities like the book exchange they hold one Sunday a month. A name would help these efforts.

People argued for various designations, but in the end the common usage of *sans-nom* (the nameless place) prevailed. Accordingly, the intersection was recently elevated to the rank of place and for the first time given a name, Place Sans-Nom, the first such appellation in France.

However, only half of the intersection is Place Sans-Nom; the other half actually remains nameless, allowing businesses to retain their current addresses.[14] Just imagine the police report for a fender bender: One vehicle might have entered Nameless Rotary just as the name was about to switch to being the nameless rotary, while another already in the nameless rotary was changing lanes for Nameless Rotary.

In Paris, the fact that one-half of a small intersection remains nameless is notable. Streets and other road landmarks need to be identified, and they need to contribute to the map of Paris. But Paris is unusual. Other places in France follow the lead of Paris, but not to the same extent. Many cities in other countries have functional names without the narrative flair of Paris. Think of the numbered streets in nearly every American city. They help with orientation but not with learning the culture.

In most of the world, people do not use street names at all. Outside of Europe and North America, and well-traveled large cities, many people live where no streets have names, or even are even clearly identifiable. Even where streets have names, there may be no street signs or maps to indicate where an address can be found. Longtime residents may not be able to name their own street. Moreover, daily practice often avoids using what names there are. Rather than saying that Samya Cafe is "on Kwalakhu Road (no number)," a guide might say, "Toward the river from Swotha Narayan Temple, first right, then on your left."

The Role of Interpretation

Not every visitor to Paris falls in love with the street signs. Some ignore them entirely or use them only as needed when their phone hasn't been charged. Others see them as a charming aspect of the ambience but don't pause to read the descriptions or to think about why a street is placed where it is. Each visitor or resident interprets the city and its street signs in his own way based on purposes and prior experiences. Some observers may see much more in the street signs than has been indicated here.

Priscilla Ferguson speculates about these interpretations and the pedagogical outcome:

> We may well wonder about the effectiveness of this history lesson. How many passers-by actually make the associations intended? As for any other text, authorial intent comprises but one of many possible interpretations. Historical connections must be strengthened; a street name in isolation will quickly lose its effect.[15]

Without a pause to read the description and think about the location of the street, the casual visitor might not appreciate rue Rambuteau. It is one of the wide roads in Paris and, as such, is useful for vehicle transportation but dangerous for those crossing on foot. It was widened in 1839, by Claude-Philibert Barthelot, comte de Rambuteau, prefect of the former Département of the Seine. Today its name honors his work.

Rambuteau was enamored of theories of the hygienists. An epidemic of cholera had devastated Paris shortly before his appointment. He thought that the narrow, winding streets in the center were conducive to the spread of the disease.

Accordingly, he began the process of sanitizing medieval Paris. This led him first to cut thirteen-meter-wide roads. The first one, which is now rue Rambuteau, eliminated six medieval streets and dozens of buildings. He modernized the sewers and ordered the construction of many fountains, some of which still function. He developed gas lighting and planted trees along the avenues.

Neighborhoods were furnished with sidewalks, lights, benches, trees, drinking fountains, and a miracle of hygiene, the famous *pissoirs* (public urinals).[16] Under his administration, the Arc de Triomphe was completed and the construction of the great Avenue des Champs-Élysées began. Rambuteau's motto was "water, air, shade."

All of Paris, not simply rue Rambuteau, would appear quite different today without the legacy of Rambuteau's work. His street is like a map legend pointing the visitor toward the history that shapes his or her experience of the city. Among his many accomplishments, Rambuteau instituted the blue enamel street signs that instantly identify Paris as Paris.

Nevertheless, one can still work, visit friends or family, or simply enjoy Paris without reading street signs or having any clue who Rambuteau was. We each create our own space that we call Paris. The map embodied in the street signs is something that the user has to create. She must be aware of its components, but then go beyond that to see it as a guide to the city, its past and current life.

This constructive process is just part of what we do in any city. Social processes create the space itself that we call the city, we further create it in our minds as we interpret its features, and we create the map that we use to navigate it. This is the case even when we imagine that we are using a packaged commercial map, for instance, the online Google map. We each interpret in a distinctive way based on our prior experience and purpose.

SOCIAL SPACE AS SOCIAL PRODUCT

Venice is a city as iconic as Paris. Appearing in countless books, paintings, movies, and TV shows, the most casual reader or viewer recognizes the space. The gondolas, canals, bridges, buildings close to the water, medieval architecture, and encroaching sea make it instantly familiar. Henri Lefebvre argues that we are not simply presented with that Venice. It is a created space, much as are the copious depictions of Venice in the arts:[17]

> Consider the case of a city—a space which is fashioned, shaped, and invested by social activities during a finite historical period. Is this city a *work* or a *product*? Take Venice, for instance. If we define works as unique, original and primordial, as occupying a space yet associated with a particular time, a time of maturity between rise and decline, then Venice can only be described as a work. It is a space just as highly expressive and significant, just as unique and unified as a painting or a sculpture. . . . [One need not] be a "connoisseur" in order to experience Venice as festival.

Venice has an architectural and monumental unity, expressing from the sixteenth century on a unitary code or common language. In that sense it appears to be a *work of art*. However, it differs in a key way from, say, Mann's *Death in Venice*. That novella has a specific creator (Thomas Mann) and creation date (1912). But no one created Venice in that way. Although it is a unique space, it was not planned in advance.

Instead, Lefebvre argues, Venice, or any *social space*, is a *social product*, not a work.[18] Space in this sense does not exist a priori, nor is it imagined and created like a painting. Instead, it is produced by the society based on its means of production:

> Every society . . . produces a space, its own space. The city of the ancient world cannot be understood as a collection of people and things in space. . . . [It] had its own spatial practice: it forged its own—*appropriated*—space . . . with its own specific time or times (the rhythm of daily life), and its particular centers and polycentrism (agora, temple, stadium, etc.) . . . each society offers up its own peculiar space, as it were, as an "object" for analysis and overall theoretical explication. I say each society, but it would be more accurate to say each mode of production.[19]

Lefebvre describes how a city's economy, expressed through its modes of production, creates the urban space. This process may take centuries. The result is then layers, remnants of past modes, and an ongoing reconstruction as new industrial processes impact the city. He emphasizes an inexorable relation between the urban space so produced and the economic activities underlying it. This is why he distinguishes between Venice as a social product and Venice as a work of art.

At the same time, hermeneutic processes are involved in both the creation of that social product and its interpretation by later viewers. In late nineteenth-century Paris, new worker housing appeared to meet the needs of the growing industrial age. *Cités ouvrières* (company towns), inspired by Charles Fourier's Phalanstères, reflected the need for concentrations of workers in new factories.

The city as social product is also shaped by conceptions about what worker housing was, how commodious it should be, its architectural style, and more. These are processes of interpretation. The viewer of Paris, or Venice, today also interprets what he sees in light of his experience and purposes. Thus, cultural history merges with political economy.

BOTH WORK AND PRODUCT

Paris may be anomalous. As with any city, everyday life and the means of production (agrarian society, the feudal system, the pre-Enlightenment royalty, industrial age, etc.) continually shape its character. Every factory calls for worker housing or transportation; every footstep on a cobblestone smooths it a little; and every demand for good bread increases the pressure for another boulangerie. The space of Paris is thus a product of the society.

However, Paris, more than most cities, reflects conscious, carefully orchestrated planning as well. In this sense it is also a work, not just a product. Georges-Eugene Haussman's renovation of Paris was a vast public works program that Emperor Napoléon III commissioned between 1853 and 1870. It vastly extended the work of Rambuteau, including the clearing of such medieval neighborhoods as the Marais, wide avenues, new parks and squares, annexation of suburbs, and the construction of sewers, fountains, and aqueducts. Haussman proudly proclaimed the "gutting of Paris."

Conscious urban development has continued since then, most recently under Mayor Anne Hidalgo. Reinventing Paris recently announced winners of proposals for innovative urban projects to redevelop twenty-three sites, including old public baths, abandoned electricity substations, parking lots, abandoned hotels, empty plots of land, and brownfields. The proposed Grand Paris Express will, for the first time, provide modern public transportation between the city's vast metropolitan area and its center.

The area around the Orly airport will be transformed into an eco-area through the building of offices, conference centers, shopping malls, and parks. Meanwhile, the city is planting twenty thousand trees and creating five hundred acres of new green areas. These are all part of making Paris into a carbon-neutral city, completely powered by renewable energy. These and other projects show that the city is not simply an accidental production of the economy but a deliberate work of art.

The economic changes and deliberate reinventions by Rambuteau, Napoleon, Hausmann, Hidalgo, and others imply a city in constant change. Consider the Marais (marsh), a historic district spread across parts of the 3rd and 4th arrondissements. It has changed greatly, being an initial target for many of the urban renewal projects.

However, the Marais has maintained much of its history as well. Some things seem to endure. A survivor of all the changes is rue des Rosiers (street of rosebushes). It was built and named in 1230 and bears the same name nearly seven centuries later. It is now a favorite shopping street for high fashion, open on Sundays and holidays because of the area's Jewish population and history.

Numerous historical markers add to the possibilities for learning, albeit with specific political purposes. The prevalence of museums, historical societies, libraries, the zoo, parks, and cemeteries makes this incidental learning almost inescapable. Cities less well-endowed than Paris have similar resources for incidental learning.[20]

PARIS AS A MAP

Echoing Peter Turchi, we begin to see that interpretation of a map is really interpretation of us:

> We chart our cities, so we chart ourselves. To chart the external world is to reveal ourselves—our priorities, our interests, our desires, our fears, our biases. We believe we're mapping our knowledge, but in fact we're mapping what we want—and what we want others—to believe. . . . By "reading" a map, by studying it, we share, however temporarily, those beliefs. This explains why we can enjoy, collect, and hang on our walls maps of places we've never been and never expect to go to—even places that don't exist. Because the map takes us there.[21]

The conventional approach would be to seek a good map of Paris to find our way around. It would direct us to the appropriate street for some journey. We could locate, for example, any of the streets mentioned in this chapter. Once there, we might recognize the street name sign to confirm our destination and even to read the explanatory text. The street name sign would mark a successful end point for our journey.

But the complex of street names, including explanatory texts, all linked to historical markers, databases, websites, and other information, is also a beginning. It constitutes a map of Paris. It informs the user about neighborhood, for example, the way that the cluster of opera composers identifies the neighborhood around the Opéra in the 9th arrondissement. It indicates directions and proximity of important organizations or monuments. It is not a map on a screen or on a large sheet of paper as we normally expect, but it serves the functions of being a map of Paris nevertheless.

Many of the streets index specific physical or architectural features. Thus, they tell the user where something is in the actual world, just as a traditional map might do. *Rue Abbaye* identifies the abbey of Saint-Germain des Prés and the corresponding Place. Other "abbey" streets are named for specific abbots or abbesses. They are usually in the neighborhood of that person or their abbey.

Rue de la Gare de Reuilly, not surprisingly, takes you to the Reuilly railroad station. Rue du Bois de Boulogne is in the neighborhood of the Bois de Boulogne (Boulogne Woods) on the west side of the city. Cours de Vincennes takes you to the Bois de Vincennes in the east. These are the lungs of Paris.

Rue du Louvre runs along the east side of the Louvre art museum. Rue de l'Odéon leads directly to the Odeon theater. Rue de Bièvre runs along the ancient course of the Bièvre (Beaver) river, which is now largely underground. And so on.

Of course, as with any map, it is possible to become confused. Rue des Halles in the 1st does take you to the RER and Métro transit hub of Gare de Châtelet—Les Halles. This is a likely destination for many, because it is a major hub, connecting three of five RER commuter-rail lines and five of sixteen Métro lines. It is one of the largest underground stations in the world.

However, you might be disappointed to discover that Les Halles (The Halls) itself was demolished in 1971. It had been Paris's central fresh food market but was replaced by the Forum des Halles, a modern shopping mall built with parks aboveground and directly connected below to the massive transit hub. One needs to interpret the odonyms carefully to see whether they refer to current or historic places.

Beyond the specific signposts, we could view not only the complex of street names, but Paris itself as a map of Parisian history. The managed rivers, Seine and Bièvre, hills, parks, bridges, buildings, and more are all parts of a constructed social space. When reverse engineered, one can see processes of industrialization; political struggles, for instance, the Commune; artistic movements; royalty and republics. In contrast with cities developed in less concerted ways or exhibiting chaotic urban sprawl, Paris itself tells a coherent story. It maps its own history, and by implication, that of France, Europe—and to some extent, the entire world.

DIVERSE INTERPRETATIONS OF SPACE

The work of art known as Paris comes with labels, starting with the street signs and especially those including descriptions. Paris wants to be seen and to inform you how it should be interpreted. The signs are its exhibit labels.

Of course, an art museum visitor need not take exhibit labels as complete, nor as the only possible way to view a work. Much of the value of engaging with art comes from personal interpretation, which may go beyond the conventional account, or even contradict it.

Consider Pablo Picasso's *The Embrace*. It *depicts a nude man and woman embracing in a bedroom*. Another might focus on technical aspects, for instance, its being *painted in shades of blue and pink pastel* in the style of Henri de Toulouse-Lautrec or Edgar Degas. Another might learn from a label that Picasso was commenting on the permissive environment he found in Montmartre, where his studio was located, with public displays of affection permitted. Picasso himself said, "What I want is that my picture should evoke nothing but emotion."[22] Another viewer might share the emotion that Picasso imagined. These and many other interpretations might apply to the painting.

In a similar way, maps of Paris, also works of art, invite divergent interpretations. Some highlight cultural sites, but they can equally well be used to find a Métro station. Others may emphasize the history of Paris,or France but serve one viewer as just a foil to critique the official history. Some maps show the arrondissements clearly; others obscure the boundary lines. For some purposes, the arrondissements could be a distraction. For someone interested in Mayor Anne Hidalgo's plan to merge the 1st, 2nd, 3rd, and 4th arrondissements into one, the arrondissement boundaries would be essential. Both purposes and perceptions affect how we read a map.

Paris has so much to see that different visitors can be said to encounter different cities. The network of odonym signs shows what some city designers have construed as the "correct" path for assessing what is important and imposing order on its complexity.

PERSONAL MEANING

Everything we might say about Paris is reinforced, yet revealed in new ways, as we look at other cities. Rebecca Solnit shows this in her trilogy of atlases on San Francisco, New Orleans, and New York. Each book is a loving critique of the history and current life of the city. She writes,

> A city is a machine with innumerable parts made by the accumulation of human gestures, a colossal organism forever dying and being born, and an ongoing conflict between memory and erasure, a center for capital and for attacks on capital, a rapture. a misery, a mystery, a conspiracy, a destination and point of origin, a labyrinth in which some are lost and some find what they're looking for, an argument about how to live, and evidence that differences don't always have to be resolved, though they may grate and grind against each other for centuries.[23]

Where one visitor may focus on New York as a center for capital, another may interpret it to attack capital. Our reading of the map can be intensely personal. But that also makes it ineluctably political. Dutch colonialists may imagine (in Benedict Anderson's sense) a nation called Indonesia and a people called Indonesians, reflecting their personal motivations for power and control. The people now labeled in that way may resist this colonial imposition, in the process uniting more than they had before. Some in current times may use their own maps and labels as they seek to reconstruct their community and assert land rights.

In the sixteenth century, Timor-Leste was caught by competing imaginations. The Dutch imagined it as part of Indonesia; the Portuguese saw it as their colony, Portuguese Timor; the indigenous people had their own views but were not consulted. The Portuguese controlled the land until 1975, when it declared independence. Nine days later, it was invaded and occupied by the Indonesian military. Subsequently, it became Indonesia's twenty-seventh province. It rebelled at that, and after many years of violence and brutality, culminating in the 1991 Dili massacre, it declared independence. It was then under United Nations administration for a while. In 2002 it became fully independent and renamed Timor-Leste, using the Portuguese for "East Timor" of its colonial past.

Imagine all the competing maps during just the modern history of Timor-Leste. Through these maps, notions of nationhood, invasion, occupation, independence, and so on are reified. The maps for the United Nations, or the residents of Jakarta, Lisbon, or Dili (the capital), are undoubtedly different from one another and vary throughout time. In this way, people co-construct the social reality along with the maps that shape and describe it.

NOTES

1. Lewis Carroll, *Sylvie and Bruno Concluded*, chap. 11, "The Man in the Moon" (New York: Macmillan, 1893), 169.

2. In an example of reality imitating art, some solar geoengineers have discussed injecting aerosol particles into the stratosphere to reflect away inbound sunlight as a way to counteract global heating.

3. Hermeneutics is the study of interpretation, especially how understanding involves repeated movements between the parts of a text and the whole (the *hermeneutic circle*). It means establishing relationships between reader, text, and context. Scholars often contrast the *understanding* sought through hermeneutics with *explanation* provided through more objective approaches. A good introduction is by David Couzens Hoy, *The Critical Circle: Literature, History and Philosophical Hermeneutics* (Berkeley: University of California Press, 1992).

4. Jeff Malpas, "Hans-Georg Gadamer," in *The Stanford Encyclopedia of Philosophy*, edited by Edward N. Zalta, 2018, https://plato.stanford.edu/archives/fall2018/entries/gadamer/.

5. Jean-Claude Bouvier, *Les Noms de Rues Disent La Ville* (Paris: Christine Bonneton, 2007).

6. Priscilla Parkhurst Ferguson, "Reading City Streets," *French Review* 61, no. 3 (February 1988): 390–91.

7. France Today Editors, "Reading French Street Signs," *France Today* (blog), May 26, 2013.

8. S. A. Cunningham, "Street Signs to Teach History," *Confederate Veteran*, January 1912, 229.

9. Bernard Stèphane, *Dictionnaire Des Noms de Rues* (Paris: Mengès, 1998). Also, the Mairie de Paris website, "Recherche Des Rues de Paris," http://www.v2asp.paris.fr/commun/v2asp/v2/nomenclature_voies/Accueil.htm, allows searches for the origin of each street name, its meaning and history, length and width, arrondissement, monuments, and much other information.

10. Charles Baudelaire, *The Painter of Modern Life and Other Essays* (New York: Phaidon, 1995), 9.

11. Bruno Latour, "Visualisation and Cognition: Drawing Things Together," in *Knowledge and Society: Studies in the Sociology of Culture Past and Present*, vol. 6, edited by Elizabeth Long and Henrika Kuklick (Greenwich, CT: JAI Press, 1986), 1–40.

12. http://mathshistory.st-andrews.ac.uk/Honours/ParisNames.html.

13. Not all of the odonymns have well-understood histories. *Picpus* may come from the Latin roots *Pic* and *Pud*, both meaning "hill." Some claim that it has a Celtic origin, also meaning a little hill. These would be justified by the slight elevation of the neighborhood. Alternatively, the name may come from an outbreak of pimples that appeared in the middle of the sixteenth century. The redness, similar to insect bites, inspired the name *Pique-Puce*, later contracted to *Picpus*.

14. Denis Cosnard, "A Paris, Une Place Sans-Nom Mais Pas sans Reproche," *Le Monde*, December 6, 2019.

15. Ferguson, "Reading City Streets," 392.

16. David Downie, *Paris, Paris: Journey into the City of Light* (New York: Broadway Books, 2011).

17. Henri Lefebvre, *The Production of Space*, translated by Donald Nicholson-Smith (Oxford, UK: Blackwell, 1991), 73–74.

18. Lefebvre, *The Production of Space*, 26.

19. Lefebvre, *The Production of Space*, 31.

20. Chris Ritzo, Chaebong Nam, and Bertram C. Bruce, "Building a Strong Web: Connecting Information Spaces in Schools and Communities," *Library Trends*, September 11, 2009, 1–13.

21. Peter Turchi, *Maps of the Imagination: The Writer as Cartographer* (San Antonio, TX: Trinity University Press, 2004), 146.

22. At the the Musée de l'Orangerie, https://www.musee-orangerie.fr/en/artwork/embrace.

23. Rebecca Solnit and Joshua Jelly-Schapiro, *Nonstop Metropolis: A New York City Atlas* (Oakland: University of California Press, 2016), 1.

3

Learning

> Early maps are complex and richly textured documents that constitute a crucial element of our modern cultural heritage. We can use them to trace humanity's spatial histories, from engagements with nature to modern globalization. By studying their production, we can comprehend how art, technology, faith, and science intertwine in the human experience. By exploring how they have variously been used, we can develop a social history of culture. The study of maps thus encompasses and integrates the humanities, social sciences, and natural sciences.
>
> —Osher Map Library[1]

Greg Miller writes about maps as tools like telescopes or microscopes. They reveal hidden relationships and stories, "from the path of the Curiosity rover on Mars, to the tangle of underground fracking wells in North Dakota, to clusters of unvaccinated schoolchildren in California."[2] One example shows both the power of maps for scientific inquiry and their potential as tools for social change.

In the mid-2000s, a team at the *South Florida Sun-Sentinel* mapped FEMA (Federal Emergency Management Agency) disbursements after several hurricanes, including Hurricane Frances, which struck in 2004. Len De Groot, director of data visualization at the *Los Angeles Times*, discovered an odd fact: "To our surprise there was one zip code in Miami where we saw there was a spike in payouts in areas where we knew there was very little damage." That led to a broader investigation by the paper, which revealed widespread fraud, a Pulitzer Prize nomination, and ultimately policy changes at FEMA.

This simple story exemplifies how maps are crucial for the development of scientific inquiry, taken broadly here to designate the disciplinary pursuit of knowledge in all aspects of the physical, biological, and social realms. As Miller says, they show us things we cannot see directly. This is vital for both the general public and for science. Across virtually all fields of inquiry, maps can make data come alive. One's capacity to do science is defined, to a large extent, by the ability to read and create maps and to move between visual and conceptual modes of understanding.

An incredible variety of maps exists, as well as ways to employ them for solving problems and learning. These maps are more than vital tools for diverse practical purposes and academic disciplines. They are tools for telling the story of our exploration of the world, real or imagined. A case in point can be seen in research exploring the function of a mandala as a cognitive graphic of sacred space.[3] This is why a classic work in the field says,

> There are specific maps and general maps, maps for the historian, for the meteorologist, for the sociologist, and so on without limit. Anything that can be spatially conceived can be mapped—and probably has been. Maps range in size from those on billboards or projection screens to postage stamps, and they may be monochrome or multicolored, simple or complex. They need not be flat—a globe is a map; they need not be of earth—there are maps of Mars and the moon; or for that matter, they need not be of any—place real—there have been numerous maps made of such imaginary "places" as Utopia and even of the "Territory of Love."[4]

MAPS OF HEAVEN

Astronomy is one of the first natural sciences.[5] Could the very first map have been a diagram of the brightest stars, possibly showing Sirius, Alpha Centauri, and Vega as dots in the sand or on a cave wall? Would Orion, under some name or another, have been their guide? Using such maps, ancient astronomers could investigate the seasons, phases of the moon, and effects on weather, fishing, and crops in surprisingly sophisticated ways.

By the tenth century BCE, Assyro-Babylonians had begun systematic astronomical observations, introducing *astrometry* as one of the first branches of astronomy. It allowed ancients to track time, which was fundamental for agriculture, religion, and navigation. Related observations led to various stone and wood henges and ground maps across many cultures.

By the third century BCE, Greek astronomers spread across the Eastern Mediterranean began to use astrometry to estimate cosmic scales. Astronomy flourished in Alexandria, Egypt, with its famous library and museum. Hipparchus of Nicaea (an ancient Greek city in northwestern Anatolia) compiled the first stellar catalog. He also created the magnitude system for describing the brightness of stars, which is still in use today, and studied the relative sizes and distances of the Sun and the Moon.

Continuing work in astrometry has led Jamie Ferguson and the European Space Agency to explore new ways of processing the enormous quantity of data contained in stellar catalogs. After analyzing various catalogs of stars, both ancient and modern, Ferguson created software to map the various astronomical parameters into a soundscape. The process, called *sonification*, represents data as sounds, an aural map. The human ear is incredibly efficient at recognizing patterns and anomalies in those sounds. Sonification could be useful for any astronomer, but it also opens the field to people who are visually impaired or simply need an alternate path into dealing with complex data sets.[6]

CONSTRUCTING MAPS TO LEARN THE WORLD

When we view an online map or a beautiful printed map, it is easy to ignore the complex processes that necessarily went into making it. It involved decisions about purpose and audience, scale, orientation, symbols, and more, as well as aesthetic decisions about coloration, font, and such. But prior to all that, a constructive process was involved in conceptualizing a map in the first place.

Early cartographers had to imagine an object that would in some way represent aspects of the world they inhabited but not be the same as that world. They had to address issues of geometry, for example, going from spherical to planar representations of the Earth or choosing whether to show topographical features, and subterranean or oceanographic spaces. The conventions of mapmaking had to be invented, most of them in a context with crude tools and limited knowledge.

Flattening the Orange Peel

In 1569, the Flemish geographer and cartographer Gerardus Mercator created a cylindrical map projection of the 3-D Earth. It had the ability to represent lines of constant bearing (*rhumb lines*) as straight segments that conserve the angles with the meridians. This made it useful for navigation at sea and later by air. Navigators could use simple instruments like protractors to plot a course.

The Mercator projection later became the standard classroom map. It told countless schoolchildren that Africa and Greenland are roughly the same size, rather than Africa being fourteen times larger (figure 3.1). As many a student has asked, "Why isn't Greenland a continent?" The Mercator map portrays Antarctica as a massive structure, five or six times the size of Africa, rather than one-half the size. It is straight on three sides, totally unlike the roughly circular area around the South Pole that you see on a globe. All the areas on the globe are distorted in size and misshapen.

Many lies are told to schoolchildren, but it is unlikely that the specific distortion of the Mercator projection was intended to mislead children. For navigators, it meant that the shortest sea route could be plotted with a straight edge. A ship can sail in a constant compass direction to reach its destination

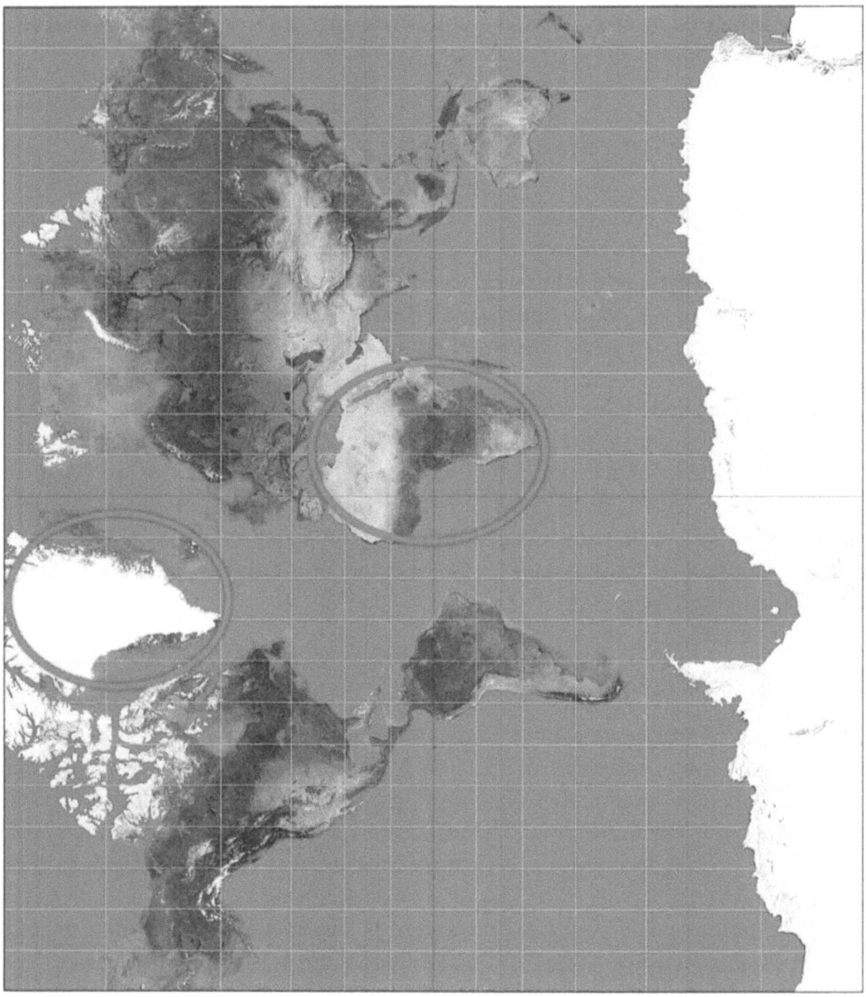

Figure 3.1. Greenland and Africa on a Mercator projection. ***Public domain***

eliminating continual course corrections that are prone to errors. However, for general use, a more accurate portrayal of the area and shape of regions is more useful.

A project that preserves area and shape (mostly) is the AuthaGraph World Map,[7] which Hajime Narukawa developed in 2016. He had worked for six years on the design, eventually finding a way to flatten the orange-peel surface of the Earth to make a 2-D map. His AuthaGraph angles continents in a way that accurately displays both their relative sizes and the distances between them.

The map is made by first dividing a spherical surface into ninety-six equal triangles. It is then transferred to a tetrahedron, maintaining the proportions of areas. It can then be unfolded to become a rectangle. It can be tiled in any direction without visible seams. From this map tiling, a new world map with a triangle, rectangle, or parallelogram outline can be formed with various regions at its center.

The map is already used in Japanese schools. A paper assembly kit mimics the way the projection itself is made. It is far more accurate for many purposes, but when laid out in any particular flat form it always introduces some distortions.

Eratosthenes

Examples of how map construction can be part of the thinking process come from efforts to measure the Earth. The most outstanding early figure was Eratosthenes, who was born circa 276 BCE in Cyrene, Libya, and died eighty-two years later in Alexandria, Egypt. He was a scientific writer, geographer, astronomer, music theorist, inventor, and poet who made the first known measurement of the size of Earth, one that proved to be far more accurate than any subsequent ones until the modern era. He was also the second librarian of the greatest library in the West, in Alexandria.[8]

Eratosthenes was intrigued by the story that on the summer solstice the sun's reflection could be seen at the bottom of a well in Syene (in Egypt, near Aswan). This had to mean that the sun was directly vertical at that location on that day. Moreover, if the Earth were spherical, then at that same moment the Sun's rays would not be vertical at another location, say Alexandria. The deviation from vertical would depend on the Earth's circumference and the distance from Syene to Alexandria. In Syene, there was no shadow; the sun shone directly down the well. But in Alexandria, there would be a shadow, so he just needed to compare an object of known height, perhaps an obelisk, with the length of its shadow (figure 3.2).

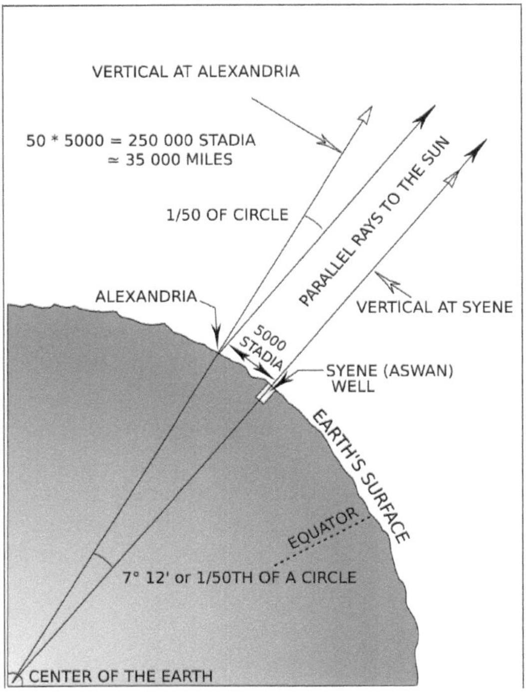

Figure 3.2. Eratosthenes method for calculating the size of the earth. ***National Geodetic Survey NOAA, public domain***

The length of the shadow told him the angle of the Sun's rays at that location. Using basic geometry, he could calculate how much of an arc on the Earth's surface lay between Syene and Alexandria. From this, the distance from Syene to Alexandria would be one-fiftieth the circumference of the Earth.[9]

Eratosthenes could not look up the mileage from Syene to Alexandria on his phone. But knowing how long camels took to travel that distance and their pace, he calculated 5,040 stadions, using an early Greek unit of length. From this, he was able to measure the circumference of the Earth very close to the current accepted value. The major uncertainty lay in the size of the stadion. Using what is about 520 modern feet, he would estimate a circumference of 24,818 miles.[10]

Eratosthenes's calculations imply his making diagrams of the angle of the Sun's rays and the shape of the Earth. He made maps to learn how to make other maps. He continually invented measurement techniques, physical devices, mathematical procedures, charts, and maps. He surely used a wide range of these, even though much of his writing has been lost. Although his work was quite sophisticated, key aspects of it can be recapitulated by children using simple tools, for instance, a globe and a measuring tape.[11]

Eratosthenes constructed many geographic and astronomical models in his work. He invented the *armillary sphere*, which is a model of objects in the sky, consisting of a spherical framework of rings, centered on Earth or the Sun. It represents the ecliptic and lines of celestial longitude and latitude. He created the first global projection of the world, incorporating parallels and meridians. The work laid the groundwork for the development of accurate globes.

He was also the first to calculate the tilt of the Earth's axis. He calculated the distance from the Earth to the Sun, which was needed at a gross level for his measurement of the Earth's circumference. He invented the leap day. He was the founder of scientific chronology, calculating the dates of important events since the conquest of Troy. He also introduced the *sieve of Eratosthenes, a* method for identifying prime numbers.

Remember that Eratosthenes did his work in the third century BCE without benefit of satellite imagery, GPS, computers, or other modern tools and theories. He made enormous contributions to our knowledge of the Earth and its position in the solar system, to the tools of cartography, to mathematics, and much more. Unfortunately, as often happens, much of his work was lost or suppressed for reasons he could not control.

Eratosthenes's measurement of the size of the Earth was a useful and truthful picture for its day. It was, of course, a limited picture; as the cartographic paradox implies, it told white lies. But much more than that, it was systematically ignored or rejected by most people, including Ptolemy in the second century CE and later by explorers such a Columbus. They preferred Ptolemy's calculation of a smaller Earth, because it made it easier to justify funding for their journeys. They chose a picture that was useful for their purposes over what is now recognized as far more truthful picture.

The once obscure Eratosthenes has justly become a modern hero, with books, children's books, documentaries, websites, and more celebrating his work. However, for our purposes we need only note that maps were an integral aspect of not only his thinking, but also that of his detractors—even those who deliberately hid his calculations, to justify the early imperialist voyages thought with maps, albeit more faulty ones. The construction and use of maps are inseparable from our cognition and our history.

Mental Models of the Earth

About thirty years ago, Stella Vosniadou and Bill Brewer conducted an experiment with elementary-school children on their knowledge about the Earth. Children in grades 1, 3, and 5 were asked a series of questions about the shape of the Earth. Many children said that the Earth is round, but they also claimed that it has an edge that people could fall over. Maps of the Earth that they would draw reflected their mental maps.

Apparently, many children were using a nonstandard mental model of the Earth. These models included a rectangular Earth, a disc Earth, a hollow sphere, and a flattened sphere. Some models were based on assumptions drawn from their everyday experience. Models like the hollow sphere or the flattened sphere were most likely generated to resolve an inconsistency between their initial model of a flat Earth and the scientific model of a spherical Earth.[12]

In the process of knowledge acquisition, children seek consistency between their direct experience and what they hear from parents and teachers. For example, when shown a photo of the Earth taken from space and told that it shows that the Earth is round, some children develop a two-Earth model: They describe first the Earth that we know and can see, which is flat, and then a second Earth that astronauts can see in outer space. Maps, whether drawn on paper or simply imagined in the head, are a crucial part of their thinking process.

CONSTRUCTING MAPS TO LEARN SOCIAL REALITY

Maps help us understand, or create an understanding of, both the natural and the social worlds (to the extent that those can be separated). For Eratosthenes, they were necessary tools to imagine how the world is shaped, how large it is, how it relates to the Sun and the Moon, and how all of that connects with daily life on Earth. In a similar way, maps both reveal and create our social reality.

Lying with a Purpose

The epigraph to this chapter suggests that lying is an unwanted but unavoidable aspect of mapmaking. But Columbus had a purpose in selecting Ptolemaic or other maps over Eratosthenes's, which went beyond technical considerations. His example extends Monmonnier's argument about the way that the cartographer's purpose causes maps to lie.

Because maps are integral to thinking, they participate in our seeking power, declaring identity, or joining communities. Appropriation, modification, and lying about maps are as much a part of the story of why maps are not totally faithful reproductions of reality as are technical compromises, for example, the one required when projecting a three-dimensional surface onto two dimensions.

Bruno Latour writes about maps as being *immutable mobiles.* This means that they are permanent (immutable) records, just as writing is, and are portable, for the purposes of sharing, problem solving, indoctrination, and other purposes. He offers an extended example based on the voyages of French explorer La Pérouse:

> La Pérouse travels through the Pacific for Louis XVI with the explicit mission of bringing back a better map. One day, landing on what he calls Sakhalin he meets with [the] Chinese [people there] and tries to learn from them whether Sakhalin is an island or a peninsula. To his great surprise the Chinese understand geography quite well. An older man stands up and draws a map of his island on the sand with the scale and the details needed by La Pérouse. Another, who is younger, sees that the rising tide will soon erase the map and picks up one of La Pérouse's notebooks to draw the map again with a pencil.[13]

Certainly differences exist in the ability to make or read maps. But in this case, that is not the salient contrast. The Chinese are quite able to think in terms of a map for navigation, probably for fishing and weather prediction. But they use maps in ways different from those of La Pérouse and therefore construct them differently.

> La Pérouse does something that is going to create an enormous difference between the Chinese and the European. What is, for the former, a drawing of no importance that the tide may erase, is for the latter the single object of his mission. . . . He is passing through all these places, in order to take something back to Versailles where many people expect his map to determine who was right and wrong about whether Sakhalin was an island, who will own this and that part of the world, and along which routes the next ships should sail.[14]

La Pérouse is not concerned with many details that might be represented in the map, perhaps not even with high precision regarding the coastline, which might be more relevant to residents of the island. He has specific purposes related to imperialism and in that sense is amenable to having maps that mislead in ways unrelated to that overriding purpose. A "better map" simply means one that answers such pertinent questions as whether Sakhalin is an island, who owns it, or which route an invading ship should take.

He is concerned with the construction of the map in the physical sense. It cannot be on sand that must remain on the beach and will be washed away by the tide. It needs to be on paper to be transportable. The scale needs to be large enough to reveal the answers to his questions. He probably lies to the Chinese about his purpose. Meanwhile, the Chinese he speaks with have other purposes. They need neither a permanent nor a mobile record because they all know the shape of the island and how to reproduce it if needed.

The story about map construction and use is not one of making as accurate a copy of the world as we can and then resigning ourselves to inevitable small, white lies. It is, instead, part and parcel of the story of our thinking and action in the world.

Center of the World

The thinking in which maps are involved does not have to be scientifically valid, virtuous, or culturally sensitive. They simply play a significant role in much that we do, for better or worse. Their contribution may not be to provide an accurate representation of physical space, but to serve purposes of cultural identity.

Consider the quest to identify "centers of the world." These centers seem to be everywhere, each signifying by its presence the yearning for a stable ground, but by their proliferation, undermining any notion of centeredness. They signify not the center of the sphere of the Earth, as Jules Verne would have it, but the purported center on the spherical surface.

The Milyon column in İstanbul is one of many monuments to "the" center. The column is all that is left of a monument that was the starting point to measure distances for all the roads leading to the cities of the Byzantine Empire. It lasted for more than a thousand years but disappeared at the start of the sixteenth century. During modern excavations, some partial fragments of it were discovered and erected again. The column served the same function as the Miliarium Aureum of Rome, another center of the world, which was displaced when Emperor Constantine I the Great remade Byzantium into his new imperial capital.

Today, the Milyon column stands near the Basilica cistern, another ancient monument, which was covered up, then rediscovered in the sixteenth century. The cistern is a huge underground room to hold water. It measures 453 feet by 212 feet, which is larger than a World Cup field.

It was built in the sixth century during the reign of the Byzantine Emperor Justinian I. During his era, and for centuries to follow, the cistern held fresh water for the citizens of Constantinople, and the Milyon dome resting on four pylons marked the center of their world. As you explore the cistern's eerily lit walkways, it is easy to imagine that you actually are in some mystical center of the Earth.

The cistern was unknown in the Western world until Petrus Gyllius discovered it while doing research on Istanbul's Byzantine remains. He was surprised to see people getting water with buckets from well holes, some within their own homes, and even catching fish!

Some calculate the geographic center as being the same location as the Giza pyramid, which would be a coincidence supporting many mystical accounts. But more commonly, it is calculated as being in the eastern Mediterranean Sea about halfway between Athens and Alexandria. Other methods locate it in northeast Turkey.

In any case, it is not too far from Istanbul. That fabulous city straddles Europe and Asia, and through its ports and the Bosporus it serves as a bridge to Africa. Its layers of civilization locate it between old and new, encompassing many religious traditions. If you had to choose one center, Istanbul and its Milyon column would not be a bad choice.

But we have many other "centers of the world." For example, Zhōngguó is the most common name for China, the first character, zhōng (中) meaning "central" or "middle," while guó (国) means "state" or "kingdom," so Zhongguo is the "Middle Kingdom" or "Central Kingdom." The name arose from the belief that the Zhou Dynasty was at the "center of the world." Tian Tan, the Temple of Heaven, served to mark the exact center spot.

The famous map (1602) by Matteo Ricci puts the "Central Kingdom" clearly at the center of the world. *Kunyu Wanguo Quantu* (Map of the Ten Thousand Countries of the Earth) is the earliest known Chinese world map printed in the style of the European maps (see the map viewer on the University of Minnesota site).[15]

Another important center is the Ka'aba ("The House of God"), a cuboid building in Mecca. It is the most sacred point within the most sacred mosque (Al-Masjid al-Haram), making it the most sacred location in Islam. Muslims face the Ka'aba to pray. For similar reasons, many see Jerusalem as the center of the world.

Easter Island (Rapa Nui) was named by the first recorded European visitor, Jacob Roggeveen. He encountered it on Easter Sunday and named it from his perspective, thus revealing how mapping so often reveals the operation of power. The phrase *Te pito o te henua* is considered by some to be the original name of the island. A French ethnologist translated the phrase romantically as "the navel of the world." This suggests that early people there saw themselves at the center and had no interest in the Christian holiday that designates their land today.

Some people suggest that a true geographic center can be calculated, but many different definitions exist. One is for a center being a *pole of inaccessibility*. This is a location that is difficult to access because it is the most distant point from the coastline. By that definition, the pole of inaccessibility in Eurasia lies in northwestern China, close to Kazakhstan. It is near the Dzungarian Gate, an important gateway between the East and the West.

Others suggest that a center of population would be the true center. That would put the center in Afghanistan.

One could also use the geographic center of landmass on the Earth. But that is subject to a variety of definitions. Antarctica is a big complicating factor because it is hard to know what of its coastline to include, especially as that changes due to global heating. Google Maps has marked the Central Anatolian province of Çorum (Turkey) as the geographic center of the surface of the Earth. Calculations by physicist Andrew J. Woods marked the center as 150 kilometers southeast of Ankara.[16]

Andrea Bianco

As representations of reality, even though sprinkled with lies, maps are endlessly fascinating and useful as tools for many purposes, not just for finding one's way. But as constructed artifacts, they embody a mix of physical reality and human passion—becoming devices for power, greed, delusion, hope, art, humor, and more. Mark Monmonnier discusses the lies of maps as unavoidable imperfections due to the need to achieve some purpose with a map. But maps can also represent more deliberate lies.

Portugal's ruler Prince Henry the Navigator sent various expeditions into the western Atlantic and along the African coast, beginning in 1418. These voyages were secret. There was a real interest in adding to the knowledge of the world, but that was coupled with a desire to use that knowledge for private gain. Not all of the discoveries were shown on published maps, and some were designed to mislead commercial rivals by concealing the existence of new lands and resources.[17]

The Portuguese had probably reached the Antilles archipelago at the eastern edge of the Caribbean by 1430. Between 1436 and 1448, Andrea Bianco made, but did not publish, maps showing the locations of Newfoundland, Florida, and Brazil. Later Portuguese maps, published in 1459 and 1489, show Asia with something like Florida, conveniently omitting South America.

Bianco developed the *Tondo e Quadro* (circle and square) method for seeing and measuring a return course. This was invaluable for repeat voyages to secure foreign resources. He collaborated with Fra Mauro, who made other detailed world maps and estimated the world's circumference within 10 percent of modern calculations.[18]

Prior to the Portuguese voyages, the famous Chinese admiral Zheng He (a Muslim) had circumnavigated the earth. A world map was published in China during the Ming Dynasty in 1418. It shows that the Ming navy had a rough knowledge of Baja California, the west coast of South America, as well as Labrador, Florida, and the Gulf of Mexico. The Chinese maps probably contributed to the Europeans' knowledge.

The Europeans also learned from Arabic science and technology. Arabs put south at the top of maps. When you face the sun in the morning, south is on the right, and right has positive associations. Also, with the sea to the south, their land was then on the top of the map. Europeans flipped the map to put north and themselves on top.

These points are supported by other historical accounts, which in sum show that fifteenth-century Europeans knew that the world was a sphere roughly twenty-four thousand miles in circumference and included large land masses between Europe and East Asia. The issue was not to "discover" America nor to prove that the world was round, but to map the details and determine who should control it.

This all suggests that most of what we had learned in school about the voyages of discovery was wrong and much less interesting than the fuller accounts available today. School textbooks tended to minimize or omit entirely any non-European contribution. That left out crucial parts of the story, including the cultural aspects of geography.

The textbooks also represented the discovery era as one of courageous, individualized pursuit of knowledge. Instead, the voyages were an essential aspect of empire building based on already extensive knowledge. Rather than enlightening an ignorant world, they were used to acquire knowledge, then deliberately mislead competitors.

Rousseau on Maps

Jean-Jacques Rousseau recognized the importance of a child's innate curiosity for learning. Accordingly, he saw the value of map construction in the process of development. Prefiguring the work of Lucy Sprague Mitchell two centuries later, he writes,

> His two first points of geography will be the city in which he dwells and his father's country house; then will come the intermediate places, then the neighboring rivers, and finally the sun's position and the way of orienting oneself by it. This is the meeting place. Let him make a map of all these things himself, a very simple map, at first formed by two objects alone, to which he adds the others little by little to the extent that he knows or estimates their distance and their position. You see already what an advantage we have procured for him in putting a compass in his eyes.[19]

Rousseau celebrates the child's development of situated world knowledge versus learning from books. He describes a father who went for a walk with his ten-year-old son after a long absence. Schoolboys were flying kites. The father asked his son to locate a kite based solely on its shadow. Without hesitation, the child said, "Over the highway."[20] Rousseau claims that this shows the value of spatial reasoning growing out of direct world experience.

The modern reader might well quibble about whether this anecdote from privileged life is useful for general education, whether the absentee father exhibits loving care, or whether the single question is a good measure of the son's learning. Nevertheless, we can still take away from the story the idea that learning should involve some combination of engagement with the natural world and academic study. It is probable that most ten-year-olds in Rousseau's mid-eighteenth century had precious little exposure to books and formal study, but living much of their time outdoors they would have done quite well on the question about the kite's shadow, whereas too many of the elite had little opportunity for direct experience of the world around them.

Mapping Modern Germany

In a college-level history class, students expressed and extended their understanding of European history through interaction with the map on Google Earth.[21] They located and bookmarked significant sites, adding a description for each.

For example, Regensburg, north of Munich, is glossed as "Deliberations of the Reichstag ambassadors in 1789 to address the rise of Jacobin ideas." Other pins linked to Wikipedia entries. In each case, students had to situate the site in relation to other towns and regions. They might notice, for example, that Thomas Mann spent part of his exile from Hitler in Prague, but primarily in Zürich.

Each entry could be read in the context of others and linked to. Thus, we learn about Hamburg being the home of the wealthy north German merchant family in *Buddenbrooks*: "The Elbe is one of the major rivers of Germany and one of the main cities of Germany (Hamburg) is on the Elbe River. Hamburg was an important city in the Hanseatic League. One of the characters in *Buddenbrooks* (*Benedix Grünlich*) is from Hamburg."

This is but one of innumerable educational applications of Google Earth. Standing alone, it provides geographic detail at many levels for any person, place, thing, or event one might want to investigate. It is also a slate on which one can write one's own emerging understanding of a topic.

THE THIRD DIMENSION

The Industrial Revolution in England led entrepreneurs to dig deep mines underground for coal and to build canals to transport it to industrial and population centers. They became aware of what lay under the surface of the land. The massive excavations throughout the country opened up the subterranean landscape as never before. This led to a new way of thinking about maps.

Adding a Subterranean Dimension to Maps

A geologist named William Smith (1769–1839) realized that a consequence of all this digging was that he could see and measure the rock layers (*strata*), essentially walking back in time. He added to that his knowledge of fossils, especially the changes he observed as they evolved during a span of many years.

Smith employed the basic principle that the oldest strata would generally be underneath the newer ones. To distinguish rock layers, geologists had already relied on color, hardness, and slope. Smith added the criterion of fossil type. Because species change throughout time, the fossils they leave behind also change.

Smith's work came long before the concept of evolution, but he understood that changes in fossil assemblages could be used to order rock layers. His discovery became known as the principle of faunal or *fossil succession*. The succession fossils provided him with a clock to date the various layers.

Smith created a *geologic map*, one that built upon a geographic map of England, Wales, and part of Scotland. The geographic maps until then showed planar surface features and, occasionally, topography. Smith was able to show the underlying strata, enabling users to go back in time to the beginning of the Cambrian era, six hundred million years ago. His was the most detailed map of geologic strata of his time and the first for an entire country.

To map rocks that lay underground, Smith had to show three dimensions of information (the surface plane plus the geologic strata on a two-dimensional surface). He had seen an agricultural map that used color coding to indicate soil type. He adopted that convention for his map of the strata. Color gradations showed where one rock formation gave way to the next. Smith supplemented the surface map with a cross-section of strata.

The most visible product of Smith's research was *A Geologic Map of England and Wales and Part of Scotland*. It measured six feet by eight feet six inches, with a scale of one inch for five miles (1:316,800). This was the first geologic map for such a large area in this detail.

> It is a map that heralded the beginning of a whole new science. It is a document that laid the groundwork for the making of great fortunes—in oil, in iron, in coal, and other countries in diamonds, tin, platinum, and silver—that were won by explorers who used such maps. It is a map that laid the foundation of a field of study the culminated in the work of Charles Darwin.[22]

In the more than 150 years that have elapsed since Smith's death, scientists have developed rock-dating techniques based on the rates of radioactive decay of atoms. Radiometric dating gives absolute dates to some kinds of rocks and organic materials, but most fossil-bearing, sedimentary rocks cannot be dated radiometrically. To estimate the ages of these rocks, geologists and paleontologists continue to rely on Smith's principle of faunal succession.

Smith's creation is not only a beautiful object, but also a vital document for industry, exploration, and the development of evolutionary theory. It expanded the very idea of what a map could be. In Winchester's account, it added the fourth dimension of time. In addition to its historical and scientific interest, Smith's map raises interesting questions about what a map can be. It highlights the role of artistic and scientific creation in making maps, but even more in enlarging the very idea of map.

Terrain Maps

The collective blind spot about the third dimension of geography prior to Smith's work is recapitulated for individual learners. Sam Brian, a social studies teacher at City College in New York, discusses this in his geography curriculum.

One day, in a fourth-grade social studies class, the study is about ancient Egypt. The teacher asks, "Why did ancient writers talk about 'the gift of the Nile?'" Using a map, students correctly infer that the river brings water to Egypt. But when she asks where the Nile's water comes from, one tentatively offers, "From the Mediterranean Sea?"

This leads to a discussion concluding that the water is salt water, which allows saltwater fishing. Brian comments,

> The confusion of the student is total. The existence of the *north-flowing* Nile, rolling down from the mountains of Ethiopia, Burundi, et al. to the south, and emptying its fresh water along the farmlands of ancient Egypt's lower kingdom, to the north, before *exiting into* the Mediterranean Sea, was inaccessible to the student, hidden by the very maps that were intended to convey the information.[23]

Going further, Brian cites the lack of experience that would offer a foundation for investigating these questions about the Nile:

> Without some concrete experience with elevation, compass direction, and river behavior, these students of ancient Egypt . . . are reduced to a kind of hopeful but random guessing with symbols that hold no meaning for them. In the absence of experience with rivers, the blue line of the Nile on the wall map may as well represent a pipe that conducts water from the kitchen sink of the Mediterranean Sea down into the receptacle of Africa.[24]

Brian's response to this problem is to use terrain models, three-dimensional models upon which water can flow downstream from hilly regions to lowlands. Students can experience directly basic properties, for instance, water always flowing downhill. Students also work with maps of terrain models, as shown in the cover photo.

In general, students progress from terrain models, which are concrete, tangible, and connected to ordinary experience, through maps of terrain models, to conventional flat maps, which are the most abstract. Thus, they are less likely to imagine islands as floating like ships, or water flowing from the sea to the interior of a country. Brian has used this technique widely both in the United States and in Nepal, which, of course, has the greatest elevation differences.

In any region, conventional maps usually obscure the third dimension of geography. At best, they may show difficult-to-interpret contour lines, heights of peaks, or other elevation markers, all of which are highly abstract for the young learner.

Comprehending the Passage of Time

Smith's work has an unavoidable conceptual problem. It is nearly impossible to comprehend the huge time span involved. Geologic periods are delineated in millions of years, but even one million years ago is before the first anatomically modern human, and the entirety of human civilization goes back only 1 percent of one million years, ten thousand years.

Smith tried to show this with fine map detail, coloring, and color gradations, but it is fair to say that even those who know the dates for all the geologic periods have difficulty comprehending them in a visceral way. An alternate approach is to turn the planar map into a physical exercise.

Geologist Robert Hazen proposes a model for a walk, in which each step represents a century back in time.[25] A century is longer than the lifetime of most humans, but it is barely noticeable on a geologic time scale.

Starting in Manhattan, one cannot even leave the building before passing all of human history. Continuing at that pace all the way to the Pacific Ocean, the walker still does not reach the early Cambrian period, when multicellular life as we know it began. And doing that would represent only 10 percent of the earth's history.

One can play the game of walking through time at various scales or speeds. For a twenty-six-mile (marathon distance) rail-to-trail walk, Hazen's pace would barely reach the appearance of grass before the end but would not see the first large mammals, much less the dinosaurs, or much else of interest. To get reptiles, fish, insects, trilobites, and so on, we would need to go much faster. A fantasy like this yields an experiential map; it could be in the form of a list, a timeline, a diagram, or simply an imaginary journey. As a walk, the walkers' muscles and joints, or thirst, would express the years traversed.

Oceanography

Visual displays can integrate conceptual information, temporal and spatial coordinates, and much more in a compact representation. In doing so they become integral to the scientific process, to public policy, and to learning, for example, about oceanography.[26]

The endangered North Atlantic right whale is not listed among mammals negatively impacted by climate change. Being highly mobile it should be able to move about to find favorable conditions. Only about four hundred are left, including fewer than one hundred breeding females. They may be extinct within thirty years.

One recent study examined relations between copepod density (a food source) and right whale sightings (a measure of whale survival).[27] However, various maps in the accompanying article suggest why that may not be so. They indicate the impact of climate change, copepod abundance for different life stages, spatial location, ocean depth, and whale sightings. The maps use words, numbers, colors, graphical conventions, and spatial relations.

Maps help scientists see patterns in this complex data set. They also help the reader understand the key findings, and, prior to that, educate the reader about the terms, relations, and reasoning that goes into producing results. Deep reading of the map becomes a way to learn about oceanography.

The Gulf of Maine has been one of the fastest warming ocean ecosystems during the past fifteen years. This has caused sudden ecosystem changes. The maps show a significant decrease in late-stage abundance of a key lipid-rich copepod starting 10 to 15 years ago. Dormancy at depth has also increased, meaning that the copepods were there in earlier stages but at deeper levels in the ocean.

Coinciding with these changes, right whales began to deviate from their typical foraging migration pattern, which tracks the copepods. One hypothesis is that a warming-driven northward shift of the copepods has driven whales into regions where protections to minimize ship strikes or fishing-gear entanglements do not yet exist. Another is that they are not adapting rapidly enough to forage well in these new habitats. More time spent foraging may lead to additional mortality.

The exact details may not seem crucial if you are not a right whale. What is important to note is that to engage with a complex topic, involving interactions among species (including humans), species life stages and foraging patterns, seasonal and climatic temperature variation, ocean depth, and more, we cannot rely on tabular data or textual descriptions. We need maps to integrate the information and guide our understanding.

FROM GARDENING TO BIODIVERSITY

Many progressive educators view pedagogy as more akin to gardening than to transmitting information. For example, Friedrich Froebel likens education to trimming a grapevine, for which the gardener needs to attentively follow the nature of the plant. He notes that "we give time and space to young plants and young animals, knowing that they then beautifully unfold." We let them rest and protect them from powerful influences. But we treat the young person as "a piece of wax, a lump of clay, from which he can mould what he will."[28]

In his dying days, Froebel urged care for his garden, asking that friends take care of his flowers and spare the weeds, because he had learned much from them.[29] Weeds taught Froebel that active hindrance or constraints for a learner could hamper their growth, but that in a natural state they reveal their pure inner life. The gardening notion led him to coin the name kindergarten, an approach that greatly influenced early childhood education around the world.[30]

Curriculum Based on Gardening and Ecosystems

Teachers at the Nisarga Batika School in Kathmandu, Nepal, fulfill and extend Froebel's ideas. The school's philosophy statement describes it as "a thriving community of learners who engage in education that is holistic, relevant and meaningful. As an experiential learning school, Nisarga Batika offers an environment where each individual looks upon the world as their classroom and values self-motivated learning as a way of life." Teachers are viewed not primarily as formal instructors but as facilitators toward goals of critical thinking, self-discovery, and creativity.

The school is a good example of place-based education. It starts with learning about the environment of the school itself, then the local community. For example, every tree on the campus is labeled by species, and student-created posters illustrate environmental awareness and nurturing. Students take multiple field trips into natural settings, learning about plants and animals, geology, and other natural phenomena as well as the local history and culture. They use maps to guide their explorations and to record their learning.

Every grade level has nature-related learning, from growing plants and caring for animals to building a vegetable garden, to studying biodiversity and climate change. Grade 3 students organize a vegetable market for the plants they grow. They sell these, learning more about the plants and their uses, about commerce and the mathematics of exchange.

A fundamental ontological point is that we are in and of nature; it is not something out there to be accessed only via a URL. Moreover, only by engaging with the inherent complexity and unpredictability can we learn how to do the same in life experiences. Henry David Thoreau reinforces this point:

> Why do you flee so soon, sir, to the theaters, lecture-rooms, and museums of the city? If you will stay here awhile I will promise you strange sights. You shall walk on water; all these brooks and rivers and ponds shall be your highway. You shall see the whole earth covered a foot or more deep with purest white crystals . . . and all the trees and stubble glittering in icy armor.[31]

Maps in the Garden

The process of designing and using gardens provides unique opportunities for problem solving, measurement, geometry, data gathering, counting, percentages, and many more aspects of mapping and mathematics. Measuring the area for a plot or collecting data regarding the growth of vegetables can become daily activities that make mapping a second-nature skill. Children can learn about area as they graph out a plot, determine how many plants they can grow and how to space them based on the distances needed for each variety.

In the Tucson area around twenty-five years ago, Marta Civil and Leslie Khan saw the value of gardens for motivating everyday mathematics. Beyond basic motivation, they saw gardens as a way to connect school activities with the children's home experiences, in which gardening was a commonly shared family activity, and to value the home life and culture of the children.[32]

Children were divided into "garden groups," with each group being responsible for its own container garden. As winter approached, they needed plastic to cover the gardens and, thus, needed to find the area of these irregularly shaped garden enclosures.

Generalizing, they discussed how to find the area of an imaginary, irregular garden enclosure. The students considered several basic strategies. One was to use such objects as cubes, tiles, or mini-pots to cover the shape. They had to address basic mapping concepts like area, perimeter, and scale. The shapes were measured with two-centimeter cubes or one-inch-square tiles. They had to cope with the different sizes, which led to different size and shapes of "gaps." This led, in turn, to consideration of the tools used to design and work with maps.

The garden connected the children and their families with the school and, in particular, with school mathematics. When the teacher, the children, and their families collaborate to develop curriculum that is grounded in their experience and knowledge, culturally relevant instruction is not only possible, but actually leads to rigorous mathematics that will help these children advance in their required school mathematics programs.

The basic principles used in a child's garden can be extended to studying the global garden called the Earth. Online maps from various organizations display maps of the world's biodiversity, the many different kinds of plants and animals in different parts of the world. Some are devoted exclusively to the world's ocean, covering a wide variety of marine species.

Nature in Miniature

Joseph Cornell designed micro-hikes as a way to help children slow down and learn about life at an ultra-local level. A micro-hike is a very short expedition guided by a four-foot string. The hiker covers the trail inch by inch on his belly, viewing such natural wonders as fungi, moss, grass blades bent by the dew, colorful beetles, worms, or spiders. They keep their eyes no more than one foot above the ground.

Cornell argues that young children are particularly fond of tiny objects, so they can exhibit an intense absorption in nature in miniature. Children should extend their strings over the most interesting ground they can find and use a magnifying glass, imagining that they are the size of an ant.

He suggests asking such questions as the following:

- What kind of world are you traveling through right now?
- Who are your nearest neighbors? Are they friendly? Do they work hard?
- What is that spider going to do—eat you, or take you for a ride?
- What would it be like to be that metallic green beetle? How does he spend his day?[33]

With a micro-hike, each inch is not a measure of time or distance but an ecozone to explore. It calls on us to use our bodies to map the phenomenon. Maps can help us become more aware of the health of natural systems, whether that is a tiny garden in a schoolyard or the entire earth.

STORIES OF INQUIRY

In its original sense, *science* (from Latin, *scientia*) meant *knowledge*, especially knowledge that people could communicate and share. Its contemporary meaning emphasizes the process of acquiring knowledge, recognizing a systematic enterprise that builds and organizes what we know in the form of testable explanations and predictions about the universe. But throughout time, and across disciplines, science has always implied producing a communal account of some problem along with possible solutions.

Both to create the story of a scientific inquiry and to relate it to others, scientists use verbal language, mathematics, and maps. Of those three, maps are often the most useful for exploring phenomena. They are the ground on which the inquiry occurs. This makes them a vital, yet underappreciated tool in the scientific process across every discipline.

Beyond the technical value of maps as tools for inquiry, they serve as ways to shape and share what we know. They are essentials of the culture, defining aspects of who we are. We can chart our growth in perspective across time and peoples through the maps we have created. They are, thus, necessary artifacts for connecting us to one another in a divided world.

Our separation from one another is mirrored by our estrangement from nature. Many living things, notably *Homo sapiens*, are under threat from climate disruption, air and water pollution, habitat loss, and the reduction of biodiversity. We have lost contact with animals, plants, and natural habitats; even the stars have disappeared. The amount of environmental degradation increases with each generation, and yet we begin to accept the only experience we know as the normal one.[34] Maps are by no means a guarantee of health in the social or natural world, but they offer us tools for the healing we need.

NOTES

1. "Osher Map Library and Smith Center for Cartographic Education, University of Southern Maine," n.d., https://oshermaps.org/about.
2. Greg Miller, "How to Tell Science Stories with Map," *Open Notebook* (blog), August 25, 2015.
3. Susan M. Walcott, "Mapping from a Different Direction: Mandala as Sacred Spatial Visualization," *Journal of Cultural Geography* 23, no. 2 (2006): 71.
4. Arthur H. Robinson and Barbara B. Petchenik, *The Nature of Maps* (Chicago: University of Chicago Press, 1976), 15.

5. European Space Agency, "A History of Astrometry—Part I, Mapping the Sky from Ancient to Pre-Modern Times," December 15, 2019, https://sci.esa.int/s/8g1qyKw.

6. European Space Agency, "From Hipparchus to Hipparcos: A Sonification of Stellar Catalogues," September 22, 2016, https://sci.esa.int/s/Welxx6w. Hipparcos was a scientific satellite of the European Space Agency, launched in 1989 and operated for five years. It was part of the first space experiment devoted to precision astrometry, the accurate measurement of the positions of celestial objects on the sky.

7. *Authalic* means equal area.

8. Some scholars claim that Eratosthene's approximation was within 2 percent or even 0.5 percent of its actual value. Others have cited 16 percent. The discrepancy arises from the lack of agreement on the unit of length—the stade. Uncertainty surrounds the length that he employed and whether he measured the variables used in the calculation himself. In any case, his estimates were much better than those of any of his contemporaries.

9. Nicolas Nicastro, *Circumference: Eratosthenes in the Ancient Quest to Measure the Globe* (New York: St. Martin's Griffin, 2008).

10. From *Geodesy for the Layman*, http://www.ngs.noaa.gov.

11. Harold Rugg and Louise Krueger, *The First Book of the Earth* (Boston: Ginn, 1936), 19.

12. Stella Vosniadou and William F. Brewer, "Mental Models of the Earth: A Study of Conceptual Change in Childhood," *Cognitive Psychology* 24, no. 4 (October 1992): 535–85.

13. Bruno Latour, "Visualisation and Cognition: Drawing Things Together," in *Knowledge and Society: Studies in the Sociology of Culture Past and Present*, vol. 6, edited by Elizabeth Long and Henrika Kuklick (Greenwich, CT: JAI Press, 1986), 5.

14. Latour, "Visualisation and Cognition," 6.

15. "Matteo Ricci, Li Zhizao, and Zhang Wentao: World Map of 1602," *University of Minnesota Libraries* (blog), n.d.

16. "Google Marks Turkey's Çorum as Center of the Earth," *Hürriyet Daily News*, June 15, 2016.

17. Frank G. Slaughter's fictionalized account of Andrea Bianco's life represents accurately much of what is known about Bianco, while filling in the gaps with a plausible story, *The Mapmaker: A Novel of the Days of Prince Henry, the Navigator* (Garden City, NJ: Doubleday, 1957).

18. David Pratt, "The Ancient Americas: Migrations, Contacts, and Atlantis," *Exploring Theosophy: The Synthesis of Science, Religion and Philosophy* (blog), August 2011, http://www.davidpratt.info/americas1.htm.

19. Jean-Jacques Rousseau, *Emile or On Education*, trans. and notes by Allan Bloom, ed. (New York: Basic Books, 1979/1762), 171.

20. Rousseau, *Emile or On Education*, 162–63.

21. Emily C. Bruce, personal communication. The site is Google Earth, Mapping Modern Germany, https://shorturl.at/bkOS4.

22. Simon Winchester, *The Map That Changed the World: William Smith and the Birth of Modern Geology* (New York: HarperCollins, 2001), xvi.

23. Sam Brian, *Geography through Terrain Models and Maps* (New York: Geography and Mapping Institute [GMI], 2019), 6–7.

24. Brian, *Geography through Terrain Models and Map*, 8.

25. Peter Brannen, "Rambling through Time," *New York Times*, January 27, 2018.

26. Oceanography in this sense is obviously related to, but not the same as, charting the oceans as chronicled in Peter Whitfield's *The Charting of the Oceans: Ten Centuries of Maritime Maps* (Rohnett Park, CA: Pomegranate Artbooks, 1996).

27. Nicholas R. Record et al., "Rapid Climate-Driven Circulation Changes Threaten Conservation of Endangered North Atlantic Right Whales," *Oceanography* 32 (June 2019), https://doi.org/10.5670/oceanog.2019.201.

28. Friedrich Froebel, *The Education of Man*, trans. W. N. Hailmann (New York: D. Appleton, 1887), 8.

29. Bertha von Marenholtz-Bülow, *Reminiscences of Friedrich Froebel*, trans. M. H. Mann (Boston: Lee and Shepard, 1892), 29.

30. Michael Steven Shapiro, *Child's Garden; The Kindergarten Movement from Froebel to Dewey* (State College: Pennsylvania State University Press, 1983).

31. *Thoreau on Nature: Sage Words on Finding Harmony with the Natural World* (New York: Skyhorse, 2015), chap. 3, para. 8.

32. Marta Civil and Leslie H. Khan, "Mathematics Instruction Developed from a Garden Theme," *Teaching Children Mathematics* 7, no. 7 (March 2001): 400–405.

33. Joseph Bharat Cornell, *Sharing Nature: Nature Awareness Activities for All Ages* (Nevada City, CA: Crystal Clarity, 2014), 46–47.

34. Progressive reduction in contact with the natural world leads to *environmental generational amnesia*. See Peter H. Kahn Jr. and S. R. Kellert, eds., *Children and Nature: Psychological, Sociocultural, and Evolutionary Investigations* (Cambridge, MA: MIT Press, 2002), 106.

4

Expressing

> The original Aran map was printed in an edition of several thousands, and one copy was worn out another could be obtained. But the present vinyl edition is unique, what will become of it remains to be seen. You are invited to walk on it, to dance on it, to write on it, to treat it as you see fit.
>
> —Tim Robinson, *Experiments on Reality*[1]

The previous chapters show how maps are vital for inquiry, for example, finding one's way to a destination or for inquiry across the disciplines. They also show how we use maps to construct the world we live in. Maps also offer opportunities to express our thoughts and feelings. They are tools for developing our expressive capacities, relating the world to our own experiences, and simply for fun.

Using maps, reading the maps that others created or making our own, involves both inner thought and active engagement with materials outside ourselves. The spirit of construction is both conceptual and material. Much of the satisfaction comes from seeing how our understanding grows along with the material object.

In 1906, John Dewey spoke to a joint meeting of people engaged in teaching either art or manual skills. For Dewey, the separation of these disciplines from one another was not a natural categorization, but an estrangement at the core of many of our problems:

> Art is not an outer product nor an outer behavior. It is an attitude of spirit, a state of mind—one which demands for its own satisfaction and fulfilling a shaping of matter to new and more significant form. To feel the meaning of what one is doing and to rejoice in that meaning, to unite in one concurrent fact the unfolding

> of the inner emotional life and the ordered development of material external conditions—that is art.[2]

Art is, in fact, a physical, even manual, act; manual work is creative and ought to be suffused with arts and higher-order thinking. The connection between these realms is through personal meaning, especially as it arises in the process of creation. Maps are art; they are media linking thought and experience in which meaning is born and expanded.

This chapter considers just a few examples of the role that maps play in creative expression, visual arts and music, free exploration, and play. Each example represents a way to open education to diverse learning styles. Through maps as art, our ability to imagine the physical world can grow, as well as our sensitivity to the perspectives and situations of other people. This can be a vital part of democratic education.

CALIFORNIA, AN ALTERNATIVE ATLAS

The *California Field Atlas* presents California's diverse ecology, topography, and history through maps and trail paintings. Obi Kaufmann goes beyond conventional conceptions of California and of road atlases to show living, connected systems. Layers of the natural world are portrayed using both science and art in terms of tectonic plates, watersheds, wildlife, and wildflower gardens. The keys explain natural phenomena.

The *Field Atlas* focuses on *emergent properties* (see more in chapter 5). It starts with a disclaimer: "This book is not full of road maps, and it won't help you if you're lost in the woods." Instead, it takes the basic idea of the atlas as a compendium of information to the next level. Focusing on the integration of living and physical systems, it "lays out the entirety of the state as a single, integrated being composed of living patterns and ancient processes. The book describes California's core systems of geography, ecology, topography and shows how those individual systems conspire together to become greater than the sum of their parts."[3]

Rather than portraying simply "geographical reality" as in the International Cartographic Association definition, his maps also imagine realities. What would California look like if the rivers had not been dammed and diverted? Could one imagine the restoration of the wetlands and their ecosystems? One of the maps does just that:

> Without dams, aqueducts, farms, and diversions, California's lush, riparian blueprint is revealed as a vast network of sheltered wetlands, vast marsh fields, coastal salt prairies, rich vernal pools, and long, unbroken alkaline basins.

> Teaming with fecund life, this primitive and pristine model supports enormous ecosystems of paradise restored. This map is an imaginative diagram, based on historical sources, of what a post-dam California might look like.[4]

The map shows twenty-eight wetlands areas that could, in principle (though not likely), be restored. One illustration is of a black-necked stilt, a bird found in wetland habitats. The North American populations have declined in the twentieth century due to conversion of habitat for human use and pollution affecting the birds and their food sources.

Kaufmann's *Field Atlas* is a beautiful, informative book. It is also a good example of (a set of) maps with an explicit purpose, both lamenting the impact of modern, industrial life on the ecosystems of California and imagining alternatives.

GAMES

The universe of games involving maps is large, complex, diverse, and like the physical universe, ever expanding. It includes outdoor games, involving rallies and geocaching. It includes board games, for instance, Diplomacy, a strategic board game set in Europe in the years leading to the Great War. Players seek to win possession of a majority of strategic cities and provinces marked as "supply centers" on the map.

Pokémon Go is an augmented reality mobile game. It uses mobile device GPS to locate, capture, battle, and train virtual creatures, called Pokémon, which appear as if they are in the player's real-world location. Players create and customize their own avatars, which are displayed on a map based on the player's geographic location. As players move about, their avatars move correspondingly within the game's map. Different Pokémon species reside in different areas of the world—for example, water-type species near water.

Some games teach geography and some use geography, real or fictional. Some games are based on Google maps, including using the satellite and street views. No medium is immune from gamers—board games, computer video games, special built simulators, text- and video-based games, virtual reality, and more.

The history of board games is long. Many of them are based on idealized or abstracted ideas of movement in a territory, for example, chess. Many involve control of that territory or navigation to find an exit or some treasure. It is not clear when geography games were first used for explicit educational purposes. In 1752, John Jefferys produced *A Journey through Europe, or the Play of Geography*, which took players on a tour in a fashion similar to the genre of travel narratives.

The German geography board game *Journey from Prague to Vienna* appeared in the late-eighteenth or early-nineteenth century, when geography was emerging as a popular topic for board games. "Its spatial orientation, colorful narration, and, it turns out, gameplay itself reveal much about how middle-class German-speaking children were situated socially and politically in the world."[5]

Geography games typically involve fanciful artwork, engaging base narratives, and challenges related to the geographic space. They make learning fun, even though, or perhaps because, they demand greater attention and problem solving than does their usual school activity.

LITERATURE

Casey Cep writes,

> Writers love maps: collecting them, creating them, and describing them. Literary cartography includes not only the literal maps that authors commission or make themselves but also the geographies they describe. The visual display of quantitative information in the digital age has made charts and maps more popular than ever, though every graphic, like every story, has a point of view.[6]

Peter Turchi links the process of writing to cartography. "To ask for a map is to say, 'tell me a story.'"[7] Like a mapmaker, the author must select and omit details, orient and move a reader, and navigate her own biases when writing. "This explains why it can be so difficult for beginning writers to embrace thorough revision—which is to say, fully embrace exploration. The desire to cling to that first path through the wilderness is both a celebration of initial discovery and fear of the vast unknown."[8]

Turchi shows that as with maps, the primary purpose of fiction is to orient us to the world. It can reveal more about who and how we are than our actual surroundings. That impulse of orientation suggests that we will never reach the end of writing or the end of exploration:

> The blank of the unwritten is the challenge we've chosen to face. We face it because, like explorers of the physical world, we want to know more about where—and why, and how—we live. We face it because we are, in some way, both inspired by and unsatisfied by what we know and what we've read.[9]

Mythology

The *Map of Mythical Creatures in Europe* depicts 213 mythical creatures (figure 4.1). It was created by students of the master's study program in cartography at Vilnius University (Lithuania), led by Giedrė Beconytė. They collected data on the geography of the creatures, their habitats and traits, and used modern techniques to create a map in the style of medieval maps showing fantastic and often dangerous creatures.[10]

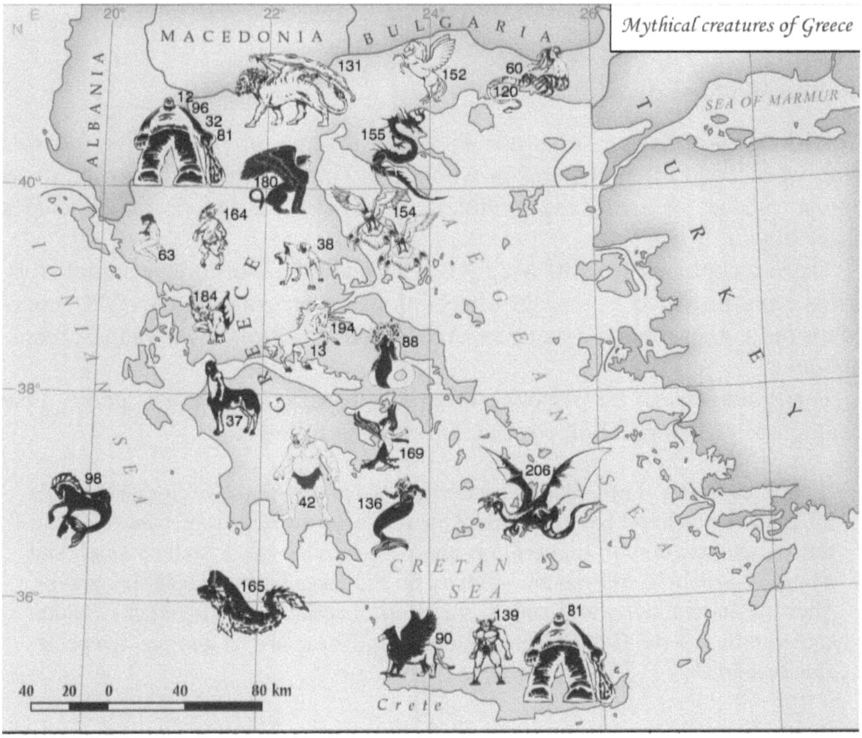

Figure 4.1. A small portion of the European Mythical Creatures maps. ***Centre for Cartography, Faculty of Natural Sciences, Vilnius University, Lithuania, 2011–2012***

The map contains original hand-drawn images of the creatures, positioned according to their location in myths. The style recalls ancient maps, but it was created using modern geographic information technology. Brief descriptions of creatures are included. The *Journal of Maps* selected it as the best map of the year in 2013.

The creators point out that mythical creatures are rapidly disappearing, so the map is the only place you can still see them. It reveals various patterns—the diversity of creatures on the British Isles, the bloodthirsty Balkans, and the differences between devils and dragons in Western and Eastern Europe. Included are trolls, centaurs, and gnomes, and also such lesser-known creatures as Adar Llwch Gwin, a giant bird that understands human language and obeys the master's orders; Aegir, an ocean giant, the king of the sea creatures; and Afanc, a Welsh lake creature that may resemble a crocodile, beaver, or dwarf. These figures live on in maps, with their geographic and cultural associations shown spatially. Some say that maps represent reality; if so, they represent many realities.

Multimedia Narrative

Traditional oral storytellers need to evoke a sense of place. They do this with words, and often gestures or whole-body movements. They point, draw hands together or apart, and modulate voices (e.g., a whisper to indicate a scary place).

These techniques are still widely used around the world. But digital story maps have provided a new set of techniques. For example, ArcGIS StoryMaps (and other tools) allow users to easily create a story with images, maps, and text.

These stories can be personal, artistic, scientific, and in any genre. One example is "Michigan Shipwrecks." It starts as follows:

> What Lies Below: The Great Lakes have provided transportation for Michigan's inhabitants for hundreds of years. Thousands of vessels from canoes to car ferries and steamers to modern ore boats have sailed these "inland seas" and unknown numbers still remain—settled on the lakes bottom in watery graves. They lie in shallow water and in the deepest reaches of Michigan's 38,000 square miles of the Great Lakes bottomlands. These are unique and irreplaceable resources.[11]

It includes an interactive map to find shipwrecks, lighthouses, boating access sites; to search for a shipwreck by name or location; to make navigation charts, imagery, or street view; and to and download data in various formats.

THE HUMAN BODY AS ART

Maps can be used to investigate and understand the human body at the macro level of organs, bones, and muscles. Diagrams in anatomy texts are considered necessary tools to display the relationships among these bodily components. But just as Leonardo Da Vinci saw the human body as inspiration, a subject for expression, and an object of study, modern artists can also map the human body in creative ways.

Jonathan Simmonds created one such map, in the style of the London Underground (figure 4.2). His map shows the arterial, venous, lymphatic, nervous, airway, and other systems, essentially summarizing important aspects of human anatomy. It has hundreds of connections and junctions, with zones, color-coded lines, stations, and interchanges. The subway features are linked with the medical terms and features.[12] Simmonds spent more than one hundred hours constructing the map. He integrated extensive research on human anatomy and the London Underground to produce a full-scale map of a human male with scientific detail, which is six feet when printed out.

Figure 4.2. Human body/London Underground map, small portion. ***Jonathan Simmonds, 2018, https://www.tuftsmedicalcenter.org***

Two-Way Maps

Simmonds's map is an example of a *two-way map*. Anyone memorizing it would have a map of the London Underground that he could carry with him always. It is literally embodied. So, his body would become a map of the Underground. Conversely, traveling the Underground would be a way of exploring the human body. The physical network of subway lines would itself be an anatomical map.

As we devise new ways of representing what we know across various domains, we discover that knowledge requires a symbolic form and that those symbols can assume ever more-diverse forms and systems of relations.

Employing processes of metaphor and metonymy familiar from literature and work with the written word, we find that the symbolic representation of say, organ structure or function, can be mapped with representations of cell types or even molecular profiles.

Mapping allows us to connect, to begin to see how our understanding of processes at one level relates to those at another. Organ function could be considered as a map of cells, and vice versa. Thus, what we consider to be a map and what we consider to be the territory can be flipped.

MUSIC

The diversity of maps reflects the fact that maps are a fundamental tool for thinking across every domain. We would not choose a map of relations between copepods and right whales to understand a sonata, but an appropriate type of map could be very useful.

Networks of Maps

Every example in this book is already operating within at least an implicit network, or it could be. For example, a map of musical structure can show the motifs, themes, sections, and movements in a sonata (figure 4.3). Map conventions can be employed to express the time for each object or the harmonic progression. Visualizations like this can be used to study the structure of an individual composition, but their power is enhanced when the map for one piece can be compared with that of another.

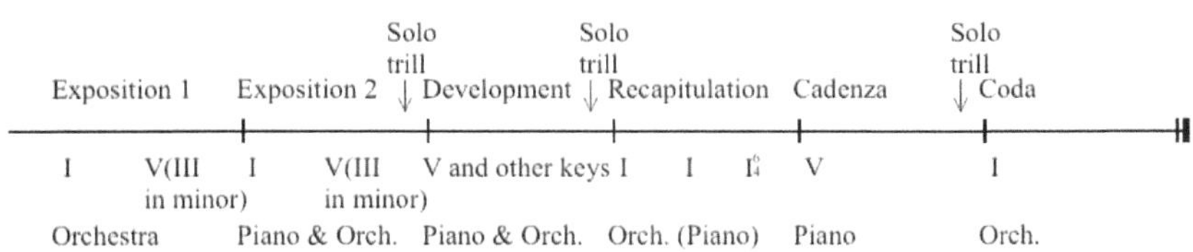

Figure 4.3. One way to map the structure of a sonata. *Wikimedia Commons, public domain*

Within a network of musical structure maps, the musicologist can explore patterns of organization throughout time. Works that deviate from the standard form have their own structures, which can be interpreted as elaborations, expansions, or contrasts with the standard form.

Beyond that, the network of maps of musical structure can be linked with timelines or concept map depictions of composers or performers. A simple biography assumes greater significance when it is linked to a composition, providing a space for speculations (or refutations of such) on relations between the music and the musician's life. The range is from simple timelines and other maps of biographies to complex, interactive timelines or concept maps for multiple composers and genres.

Some researchers have used *neural network* architecture and analytics to produce *self-organizing maps* of musical characteristics. Artificial neural networks, also called connectionist systems, are computer programs inspired by similar networks in animal brains. They learn to perform tasks after analyzing multiple examples, rather than being programmed directly for the task.

A neural network can discern multidimensional characteristics of a data set and project those onto a two-dimensional space. This has been done for sound samples from Johann Sebastian Bach, Ludwig van Beethoven, and Frédéric Chopin, with the network identifying similarities and differences among the compositions, potentially some relations that are not immediately evident to a student, or even a musical expert.

In addition, musical compositions and composers, as well as performers, producers, patrons, and other participants, have been represented on a wide variety of ordinary or fanciful geographic maps, showing the location of a composer, the place where a piece was written or performed, or its topic.

These include "Classics for Kids," the "Australian Music Map," "Music by Black Composers," the "Musical Map of the United States," and many others. The connection of geographic data, timelines, structural analyses, mathematical pattern analysis, and more makes possible a much richer ecosystem for analyzing, understanding, and making music.

Representing Music and Musical History

Tillman Weyde and Jens Wismann write,

> The relations of music within itself and to its context can be represented naturally as nodes and edges in graphs, representing musical objects or general items of information and their relations. . . . Mapping software is especially useful for non-hierarchical information, e.g., analytical relations of musical parts.[13]

Maps and mapping software can be used to represent aspects of music and musical history. They are especially useful for representing objects (e.g., notes, chords, motifs, sections, themes) and their relations (e.g., similarity, succession, containment, contrast, harmonic function).

Analysis of a sonata shows the succession of the motifs in themes, themes in sections, and sections in movements. Their map shows succession relations, for example, *exposition*, *development*, *reprise*, in one color. The length of the succession edges corresponds to the time between the connected objects. Containment relations, for example, motifs within themes, are shown in another color. Graphs of this type can be used to study the relations between motifs, themes, and sections. The map can be enhanced by introducing harmonic progression (e.g., from I to V and IV chords).

Map visualizations further offer the opportunity to use different forms of visual information display, for example, scores, images, and sound. They can include conventional geographic maps to show the location of a composer, performer, or performance. Musical ethnographers might need to show the locations of different instruments or styles throughout time.

One can see musical scores themselves as maps of music. The Western system used widely today maps the temporal and spatial relations of a piece of music (figure 4.4). For example, notes higher on the page for a given measure indicate keys on a piano in the higher tonal range, to the pianist's right. The time values for notes and dynamic markings indicate how long notes are to be held. This system is as much a map of otherwise mysterious territory as a cartographic map of a forest might be.

The same information can be indicated in many other ways. Earlier forms are found in a cuneiform tablet that was created at Nippur in about 2000 BCE. It has instructions for performing music composed in harmonies of thirds.

Various systems of musical notation have been in used in China for more than two thousand years. Musicians there (and, to some extent, in other countries) use numbered musical notation (figure 4.5). Numbers 1 to 7 represent the scale degrees (musical notes) as with solfège (do, re, mi). Dots above or below a musical note raise or lower it to other octaves.

Figure 4.4. Western-style musical notation. ***Wikimedia Commons, public domain***

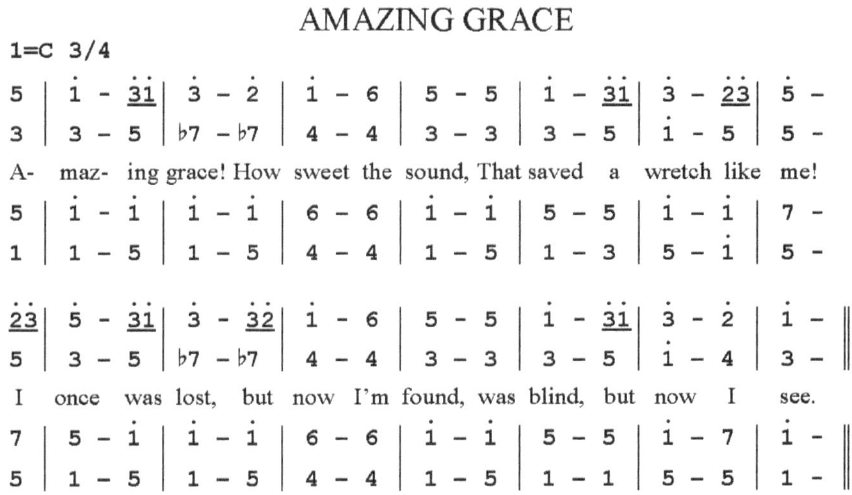

Figure 4.5. Chinese numbered musical notation. ***Wikimedia Commons, public domain***

Many systems exist, too, for representing choreography. These show body position and movement as well as physical location in a dance space.

FORENSICS

Maps have become a staple in popular media, reflecting their use in real life. They have a clear value as narrative devices, especially the way they make a setting visible to both characters and audience. Even if we discount a little for their narrative value, their prevalence is remarkable. This is especially the case for crime dramas, sci-fi, and adventure stories.

This can be seen in a randomly chosen episode ("Noisy Edge") of *NUMB3RS*, an American crime drama television series that ran on CBS between 2005 and 2010. The series follows FBI Special Agent Don Eppes and his brother Charlie Eppes, a college mathematics professor and prodigy, who helps Don solve crimes for the FBI.

The episode begins with maps shown on computer monitors behind the main characters. News arrives that the Federal Aviation Administration has received seven reports in ten minutes of a UAE (unexplained aerial event) on a path toward Los Angeles. It has no reported flight plan, and it is not visible on either civilian or military radar. Two aircraft are sent to intercept it, but they cannot spot it.

Using a set of visual sightings and a map, Charlie is able to plot a likely trajectory showing that the UAE is heading toward downtown. But the UAE still cannot be located. The next day, the team resorts to a technique called "squish-squash," which can sometimes pull a signal out of noisy data (the noisy edge). They look for a weak signal and squish and squash it mathematically to find its strength and direction. With this, they are able to estimate a new flight path for the object.

Later, reaching another dead end, they decide to focus on the object rather than the point of origin. They overlay the various radar maps to try to build an image of the object. After ruling out airfields that it might have used, they come to the conclusion that the craft uses vertical takeoff and is made of nonmetallic materials (making it difficult to detect on radar). Using maps and projecting from the new data, they calculate that the craft was heading toward a large arena where thousands of people could have been a target.

Radar screen maps and city maps continue to be used throughout the episode. Eventually, of course, their use of mathematics and maps saves the day. In other episodes, the FBI team uses techniques similar to those that John Snow employed in the nineteenth century to locate the source of a cholera epidemic. These tools of trajectory tracking, cluster analysis, and a combination of intuitive and mathematical investigation are used across many dramas and in real-life forensics.

FROM PHYSICAL TO DIGITAL AND BACK AGAIN

People typically identify the center of their city by some important building, for instance, a temple or town hall. Oftentimes, that center is known just as common knowledge, an outdoor market or viewpoint. But in many countries, it is marked explicitly, both in terms of geocoordinates and with some sort of physical marker.

That center may be kilometer zero (km 0), a particular location from which all distances are supposed to be measured. For example, the Milliarium Aureum ("Golden Milestone") of the Roman Empire was likely the literal origin for the claim that "all roads lead to Rome." Just outside Notre Dame Cathedral in Paris, a small plaque in the pavement marks the spot to judge the distance from Paris to all other places in France. Paris Point Zero has itself become a tourist attraction used to mark one's journey through France or to spark romance.

German artist Aram Bartholl is intrigued by the red map marker that Google Maps uses to mark the center of cities. This leads to questions regarding the relation of the digital information space to everyday life in the public city space. He has placed six-meter-tall wooden sculptures in the shape of Google Maps bookmarks in the purported center of various cities, including San Francisco, Taipei, Berlin, and Szczecin. He writes,

> With a small graphic icon Google marks search results in the map interface. The design of the virtual map pin seems to be derived from a physical map needle. On one hand the marker and information speech bubble next to it cast a shadow on the digital map as if they were physical objects. When the map is switched to satellite mode it seems that they become part of the city. On the other hand it is a simple 20 px graphic icon which stays always at the same size on the computer screen. The size of the life size red marker in physical space corresponds to the size of a marker in the web interface in maximal zoom factor of the map.[14]

Bartholl elaborates: "The Google map marker piece should make people think about tech-society-privacy relations. 'Map' in a way symbolizes the massive position of Google's gate to local filtered information and its influence on our perception of the city."[15]

Bartholl's move from physical to digital and back again is reminiscent of the idea of *earth sandwich*. This is created when two slices of bread are simultaneously placed on opposite sides of the Earth.

In May 2006, Ze Frank issued a challenge to his video blog audience to create the first Earth sandwich. Viewers could use an online tool to locate their antipode on the globe. Hundreds of people around the globe began sending pictures of bread on the ground in tribute to the effort. The first Earth sandwich was completed in less than one month. A baguette was placed in Spain, near Madrid, at the same time that another baguette was placed in New Zealand outside of Auckland.

Many of the examples in this chapter exhibit a playfulness. These are aesthetic responses to things we learn about the world and to the technologies that enable that learning. Their playfulness may obscure how much learning occurs when we experience these artworks.

One small example is that any United States-based fan of Frank's video about Earth sandwiches could quickly discover that her country is across from the Indian Ocean, not from any land mass. The closest land a North American viewer might find is Diego Garcia in the British Indian Ocean Territory, which is opposite a location in Mexico. The person standing at the antipode of a place in the United States would not only be looking at a different hemisphere of stars; she would be treading water.

INTEGRATING INTERNAL LIFE AND MATERIAL CONDITIONS

Mapping encourages and supports play. It allows us to express beauty, curiosity, delight, fears, outrage, and hopes. Maps can simultaneously be practical tools, cognitive creations, or works of art.

As we explore the various manifestations and uses of maps, we learn more about maps per se, expand our understanding of geography, venture deeper into both the physical and the social worlds, and encourage the unfolding of our inner life. Recognizing this should encourage us to demand education in which learners are actively engaged in making meaning for themselves, expressing their experiences through multiple media.

Artworks also expand our critical awareness. For example, Bartholl's map markers cause us to question the relation of the digital information space to everyday life. Other maps both reveal and express other relations with the natural world. The use of maps in games, literature, and other areas can help us appreciate the importance of physical space, culture, and history in all aspects of our lives. It can be a major factor in enlarging civic intelligence and promoting democratic society.

NOTES

1. Tim Robinson, *Experiments on Reality* (London: Penguin Ireland, 2019), 118.
2. John Dewey, "Essays on the New Empiricism," MW 3:292.
3. Obi Kaufmann, *The California Field Atlas* (Berkeley, CA: Heyday, 2017), xvi.
4. Kaufmann, *California Field Atlas*, 105.
5. Emily Bruce and Elise Klarenbeek, "Playing on the Map: An Educational Game from the Age of Revolutions," *Journal of the History of Childhood and Youth* 13, no. 1 (2020): 9.
6. Casey Cep, "The Allure of the Map," *New Yorker*, January 22, 2014.
7. Peter Turchi, *Maps of the Imagination: The Writer as Cartographer* (San Antonio, TX: Trinity University Press, 2004), 11.
8. Turchi, *Maps of the Imagination*, 14.
9. Turchi, *Maps of the Imagination*, 41.
10. Giedrė Beconytė, Agnė Eismontaitė, and Jovita Žemaitienė, "Mythical Creatures of Europe," *Journal of Maps* 10, no. 1 (2014): 53–60.
11. https://storymaps.arcgis.com/.
12. Mark Blunden, "Surgeon Uses London Tube to Inspire Life-Size Human Body Map," *Evening Standard*, November 30, 2018; full color, zoomable version of the map, https://i.redd.it/pvic1vxwmly11.png.
13. Tillman Weyde and Jens Wissmann, "Visualization of Musical Structure with Maps," edited by R. Parncutt, A. Kessler, and F. Zimmer (Proceedings of the Conference on Interdisciplinary Musicology (CIM04), Graz, Austria, 2004).
14. Aram Bartholl, "Map," n.d., https://arambartholl.com/map/.
15. Ian Glover, "Artist Profile: Aram Bartholl," *Rhizome*, August 24, 2011, https://rhizome.org/editorial/2011/aug/24/artist-profile-aram-bartholl/.

5

Solving

> My plea is not to make this map or that map. My plea is think of geography as a laboratory pursuit and to urge all teachers to hunt with their children for source material either in the immediate world around them or wherever scientific geographic data can be found, and to invent tools for the study of the relationships inherent in these data.
>
> —Lucy Sprague Mitchell, *Young Geographers: How They Explore the World and How They Map the World*[1]

Lucy Sprague Mitchell thought that geography was a field of knowledge that children could encounter, as both makers and users of maps, if they could approach it through experience in their own lives. She saw it not as a set of facts to be learned but as a process involving exploration, discovery, and the creation of tools to express what they found.

A map can be used to answer basic questions, for example, "Where am I?" and "Where am I going?" It can answer such specialized questions as, "Where did the Curiosity rover go on Mars, and what did it see?" Many would say that practical delivery of information is the raison d'être for a map. The full range of questions a map could answer is endless.

However, a map can not only *map* the features of some place, object, or idea, it can also *map* our path to answering some question about that place, object, or idea. Each map tells a story about the explorer's engagement with the world. The power of maps to render context as a visible, manipulable object makes them into tools for understanding. They help us place given information into a network of meaning. Maps are important components of education's ecosystems.[2] They become tools for inquiry and for action in the world. The following example about John Snow's work on cholera in nineteenth-century London illustrates this well.

TRACKING CHOLERA

John Snow (1813–1858) was an English physician and a leader in the development of anesthesia and medical hygiene. But he is perhaps best known for his work in stopping a cholera outbreak in Soho, London, in 1854. His methodical investigation of cholera cases sounds mundane at first. However, it represents a major advance in epidemiology. It also shows us one way in which maps matter.

When Snow first considered the epidemic, he did not understand the mechanism of transmission. The germ theory of disease had not yet been developed, so the presence of germs was not an obvious answer to consider. Contact tracing, a term of great currency today, was not even considered.

Today we know that cholera is caused by consuming food or water contaminated by the bacterium *Vibrio cholerae*. It is most often spread through sewage entering a water supply. Often someone with the condition inadvertently pollutes the water that those nearby use. Snow could not assume Vibrio contamination, but he did question the prevailing theory that cholera was caused by "bad air" or "miasma." He saw no evidence for bad air being the cause.

Without any other answer, or any obvious test to perform, Snow decided to conduct systematic interviews with those who had contracted cholera in the Soho area. He also created detailed maps of the area, thinking that an answer might lie in the geographic proximity of cases (see figure 5.1). His maps identified roads, property boundaries, and water lines.

By talking to local residents, Snow began to think that the source of the outbreak was the public water pump on Broad Street (figure 5.2). Note the prevalence of cases on his map shown around Broad Street. He writes, "On proceeding to the spot, I found that nearly all the deaths had taken place within a short distance of the pump."[3]

Snow might have rested at that point, but his map also revealed that not all the data could be accounted for by proximity to the Broad Street pump. Ten deaths were from houses situated closer to another street pump. Further interviews showed, "In five of these cases the families of the deceased persons informed me that they always sent to the pump in Broad Street, as they preferred the water to that of the pumps which were nearer. In three other cases, the deceased were children who went to school near the pump in Broad Street." Snow's maps thus revealed patterns, anomalies, and the direction for more detailed inquiry.

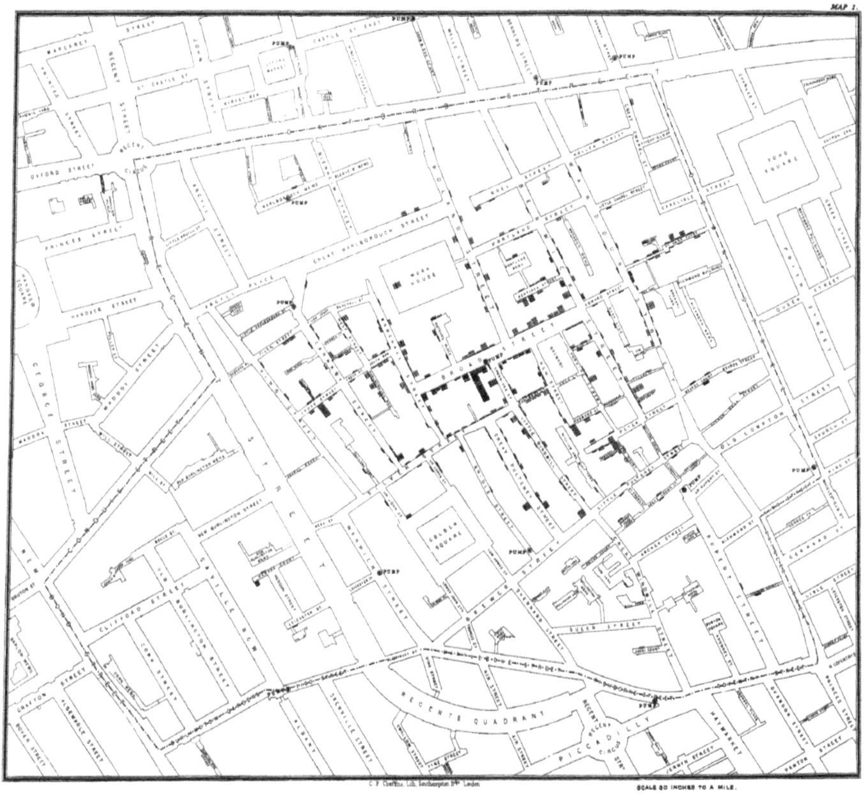

Figure 5.1. Clusters of cholera cases in the London epidemic of 1854. ***John Snow, drawn and lithographed by Charles Cheffins; public domain, Wikipedia***

Summarizing his findings, Snow writes, "With regard to the deaths occurring in the locality belonging to the pump, there were 61 instances in which I was informed that the deceased persons used to drink the pump water from Broad Street, either constantly or occasionally." Nearness to the Broad Street pump was clearly associated with the cholera deaths. Moreover, his continuing inquiry found "no particular outbreak or prevalence of cholera in this part of London except among the persons who were in the habit of drinking the water of the above-mentioned pump well." This was strong evidence that the particular pump was not only a possible cause, but the only cause.

Snow's studies of the pattern of transmission were enough to persuade the local council to remove the pump handle, thus disabling it. The handle-less pump stands today in London as a monument to his work.

Later research showed that the public well was only three feet from an abandoned cesspit, of the kind commonly found under houses at that time.

Figure 5.2. John Snow memorial and public house on Broadwick (formerly Broad) Street, Soho, London. *Wikimedia Commons, photographer: Justin Cormack, CC BY-SA 2.0*

A baby who had contracted cholera elsewhere lost a cloth diaper, which then washed into this cesspit.

Later, Snow used a dot map to illustrate the cluster of cholera cases around the pump, combining a map of cholera cases with one of streets and pumps. His use of that method of displaying data was instrumental in establishing dot maps as a major tool for cartographers.[4] The visual comparison of the representations of roads, water supplies, and cholera deaths revealed crucial information not evident from one map alone or from a simple list or table. The source of the cholera outbreak was thus a simple case of a property emerging from the context that the map provides.

Snow also used statistics to illustrate the connection between the quality of the water source and cholera cases. With these tools he was able to trace the outbreak to the Southwark and Vauxhall Waterworks Company, which was obtaining water from polluted sections of the Thames.[5]

Snow's action has been commonly credited as ending the outbreak, but Snow himself observed that the epidemic may have diminished due to people leaving the area or from some other cause reducing the cholera poison. Nevertheless, his study was a major event in the history of public health and geography. It bolstered his argument for the use of scientific methods in public health. His investigation was a key event in the development of both epidemiology and the digital geographic information systems of today.

CHARLES BOOTH'S POVERTY MAPS

The quantitative analysis of society began in the mid-nineteenth century. Employing the developing techniques of surveys and statistical analysis and using methods similar to those of John Snow, Charles Booth began a major survey of London life and labor in 1886. His work, along with such colleagues as his cousin, Beatrice Webb, inaugurated the systematic study of poverty in Britain.

Booth's poverty maps revealed a spatial component to poverty as well as an environmental context. Before that, environmental explanations of poverty mainly interested health professionals. It countered the prevailing theory that moral degeneracy accounted for poverty.

Booth's seventeen-volume study *Inquiry into Life and Labour in London* (1886–1903) is a founding text of British sociology, drawing on both quantitative and qualitative methods, including ethnography. It influenced the "Chicago School" of sociology, the discipline of community studies, and the *Hull-House Maps and Papers*.

A website, Charles Booth's London,[6] enables the user to search a catalog of more than 450 original notebooks from *Inquiry into Life and Labour in London*, view forty-one digitized notebooks, and explore the London poverty maps. Interestingly, Booth's nineteenth-century maps are best viewed today in interactive digital format, which enables zooming, scrolling, search, and the application of statistical measures. They also depend heavily on subtle color variations, showing for example, different levels of poverty.

HULL-HOUSE MAPS AND PAPERS

After Julia Lathrop began investigating applicants for relief in the neighborhood of Hull-House, the settlement house in Chicago, she realized that more detailed analysis was needed across a variety of social issues. This inspired *Hull-House Maps and Papers*, a set of projects that built upon the tradition that John Snow and Charles Booth started. Hull-House maps even retained the colors that Charles Booth used for his wage maps of London.

The Territory of Hull-House

Agnes Sinclair Holbrook describes the geographic territory surrounding Hull-House, which would be covered in the research as follows:

This third of a square mile includes east of the river a criminal district which ranks as one of the most openly and flagrantly vicious in the civilized world, and west of the same stream the poorest, and probably the most crowded section of Chicago. At the extreme northwest of the whole, on Halsted, near Polk, is situated Hull-House, within easy walking distance of the densely populated network of streets and alleys on the west side, claiming our chief attention.

It is customary for the lower floor of the rear houses to be used as a stable and outhouse, while the upper rooms serve entire families as a place for eating, sleeping, being born, and dying. Where there are alleys the refuse and manure are sometimes removed; where there are none, it would seem they accumulate undisturbed.[7]

Finding Patterns

Hull-House residents measured and mapped patterns associated with immigrants, working conditions, labor unions, social settlements, and the function of art. As Mary Jo Deegan points out, "women's moral agency was central to their use of social science to improve democracy and the lives of the disenfranchised."[8] During the next forty years, the residents studied occupations, family size, housing, milk quality, food use, and epidemiology. They displayed their findings on the walls of Hull-House for the neighbors to see and discuss, not simply report them in some academic forum.

The Hull-House maps revealed to local people their own lifestyle patterns and helped them make informed decisions about community issues. This led to efforts to establish the eight-hour day, the minimum wage, and the elimination of child labor. They worked to promote social movements, labor unions, women's suffrage, and arts and crafts.

The research took place during the span of several years under the auspices of various projects directed mostly by Florence Kelley.[9] A major portion was collected as part of a national study of slums commissioned by U.S. Congress. It showed how multiple, coordinated maps could reveal patterns not otherwise observable.

Like Snow, Kelley and her assistants visited homes and sweatshops, asking the residents questions about their ethnic origins, the number of people in the household, the wages that the principal wage earner earned, and how many weeks the principal wage earner had been employed. Data were recorded on data collection instruments, or questionnaires, then called "schedules." The information was copied every night by Hull-House residents before the schedules were sent to Washington, DC.

The residents then created maps (see a small portion of one map in figure 5.3) to display the data from the surveys.[10] The maps conveyed the data in visual, more easily understandable formats. They also helped to convince municipal authorities to reform policies. And they revealed emergent properties of the data, for example, relationships between employment and health. Their beauty and functionality have made them classics.

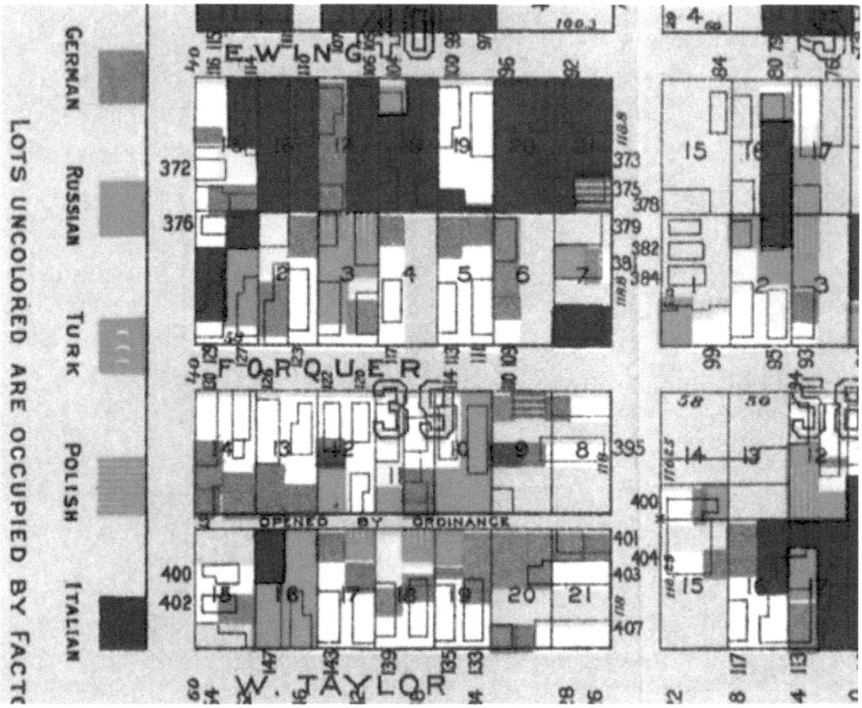

Figure 5.3. Small portion of an ethnicity map from *Hull-House Maps and Papers*, shown here in grayscale. *Residents of Hull-House*. Hull-House Maps and Papers: A Presentation of Nationalities and Wages in a Congested District of Chicago *(New York: Thomas Y. Crowell & Co., 1895).*

A good example of their usefulness is the study of "wage-earning children" by Florence Kelley and Aleina Stevens. They found that far from helping children develop valuable trade skills, industry simply exploited their cheap labor. Children moved from one place to another and rarely learned useful trades:

> Nothing in our work has been more of a revelation than the migratory method pursued by the children, which forever disposes of the only argument in favor of child-labor that before seemed valid; namely that the work afforded a sort of industrial education for the boy or girl must depend upon manual labor for livelihood in the adult years. . . .

> It is not a trade that is learned in the great workshops where child-labor is the foundation of the company's riches. What the child does learn his instability, unthrift, trifling with opportunity.[11]

Their work employed charts to document the harmful effects of factory work on children. For example, they compared the relative growth in stature between children in school and those children consigned to factory work. Their substantiation of the detrimental effects of child labor supported passage of child labor laws.

The Publications

The book *Hull-House Maps and Papers*, published in 1895, is a set of essays and statistical information that Kelley and colleagues collected. The commentaries focus on several different topics and communities, providing comprehensible narratives of how people lived and worked in the worst slums in Chicago.

In the preface, Jane Addams described the work as being simply "immediate" observations for "constructive work." This is overly modest; the effort is increasingly recognized as a pioneering achievement linking sociological research with social work and even providing a blueprint for digital geographic information systems:

> The residents of Hull-House offer these maps and papers to the public, not as exhaustive treatises, but as recorded observations which may possibly be of value, because they are immediate, and the result of long acquaintance. . . . [They] have been chiefly directed, not towards sociological investigation, but to constructive work.[12]

Despite Addams's disclaimers, Hull-House established through its statistical research a professional expertise while creating a platform to redefine poverty. It conceived poverty as a social and organizational problem, not one of moral failure. The challenge was to ameliorate the social dysfunction of American modernization.

Impact

The various projects reported in the book included the sweating system, wage-earning children, cloak makers, the Chicago ghetto, the Bohemian people, the Italian colony, Cook County charities, art and labor, and the settlement as a factor in the labor movement. The projects made extensive use of maps, and as noted in the quotation from Holbrook, awareness of the spatial relations within the neighborhood and the housing units. Long before the era of digital geographic information systems, they showed the power of map overlays to reveal emergent properties. Beyond that, the mapping projects have done the following:

- Laid the groundwork for the first court case anywhere in the world to be decided on the basis of social science data[13]
- Became the model for studies of African American communities, for example, the work of Isabel Eaton and W. E. B. Du Bois on *The Philadelphia Negro* (1899)
- Set the stage for Kenneth Clark's crucial experiments spelling the end of U.S. school segregation in *Brown v. Board of Education*
- Led to the first laws regulating child labor and women in the workplace
- Showed a model for the new field of sociology and, more specifically, for the Chicago School of sociology[14]
- Became a beautiful set of visual artifacts inspiring mapmakers ever since

THE DIGITAL AGE

The map-based projects of Snow, Booth, Kelley, and others during the nineteenth and early twentieth centuries stand well on their own. They addressed complex problems in such areas as health and economic justice, making substantial progress. Their works inaugurated entire fields of study, including sociology, social work, epidemiology, public health, applied statistics, and geographic information systems. As such, they serve as introductions for learning in any of these fields. Finally, the maps they produced are works of art, worthy of marvel today.

However, a natural question arises: How might modern mapping technologies, including overlays, zooming, browse and search, aerial and satellite imaging, statistical analysis, and shareability have been used on the same problems? Would the new tools lead to different results?

Fortunately, we can get some idea of this from the fact that many of the historical maps have been digitized, not simply as images, but as interactive sites. For example, John Snow's map tracking cholera in Soho is available online at ArcGIS. The data is taken directly from Snow's map. It starts with a layer showing Snow's original map and employs his dot map representation scheme. However, it uses dot size to indicate the number of deaths at a particular location.

The online map is designed for use in a workshop on the use of Geographic Information Systems (GIS), taught at Yale University. Measures of *spatial central tendency* have been applied to the data set. For example, one is the *spatial mean*, or geographic center of the distribution of deaths. The online map shows that lying within thirty-five meters of the Broad Street Pump. It also shows with a red (here, gray) circle the *standard distance*. This is the radius (140 meters) around the spatial mean containing 68 percent (or one standard deviation) of the deaths in the outbreak.

Snow would have appreciated these tools. They would have aided his research and the communication of the results to a skeptical audience. But it is worth noting that his hybrid qualitative/quantitative analysis may be more compelling.

He points out that "there were only ten deaths in houses situated decidedly nearer to another street-pump." However, "in five of these cases the families . . . always sent to the pump in Broad Street, as they preferred the water to that of the pumps which were nearer. In three other cases, the deceased were children who went to school near the pump in Broad Street." These observations are not captured by the standard measures of central tendency, nor in either the paper or online map display. The most effective approach is a combination, or network, of representations and tools.

Roger Tomlinson is widely recognized as the founder of (digital) Geographic Information Systems (GIS).[15] His work led Canada to adopt a layer system for maps. The Canadian Land Inventory used soil, drainage, and climate characteristics to assess the use of land for various crops or forestry. These *map overlays* are a clear example of emergence. Layering had been implicit in the earlier work of Snow et al., but computers made the process much easier and more generalizable.

Another example of the approach comes from the Penn State Environmental Resources Research Institute. It overlaid maps showing watershed boundaries, the slope of the terrain, soil types, land use patterns, and animal loading (livestock wastes). With this, they could estimate and display visually the agricultural pollution potential of every major watershed in the state.[16] The detail is not obvious from a paper map. However, with computer-based GIS one can see the variable values in color, zoom in to particular locations, compare regions, apply statistical analyses, assess the effects of selected layers, and other such manipulations.

Soon after the Canadian work, the U.S. Census Bureau adopted GIS for the 1970 census, digitizing boundaries, roads, and urban areas. New digital tools supported such features as *choropleth mapping*, in which areas are shaded or patterned according to values of a variable like population density or per-capita income. The UK Ordnance Survey also incorporated GIS into its topographic map development.

Today, these techniques are expanding to developing economies, largely through the use of OpenStreetMap (OSM), a leading example of volunteered geographic information.[17] OSM is a collaborative project to create a free, editable map of the entire world. Its rationale has been current and potential restrictions on the accessibility of map data that commercial or state interests own. Inexpensive, easy-to-use GPS devices and wider availability of internet access have facilitated its growth.

PERFORMATIVE MAPS

In language theory, *performative utterances* are those that change social reality rather than describe a preexisting reality. J. L. Austin is best known for developing this concept. He argues that some utterances "describe" or "constate" something and are, thus, always true or false. However, many others have no truth value. These include jokes, nonsensical sentences, questions, commands, moral and ethical directives.[18]

An important additional category of non-constative propositions are performatives. The uttering of a performative is an action that establishes or changes social reality. We have many examples. Some are issued by established authorities, for instance, judges or priests: "You are guilty," "You are now husband and wife," "Your name is now . . . ," "This bill is now law." But ordinary people can issue performatives as well: "We are no longer friends," "I promise to return," "My goal is to swim in one hundred ponds."

Maps are a form of utterance as well, and as such, in Austin's sense can be either constative or performative. Most are constative; they attempt to describe aspects of reality and can be judged as true in the sense of succeeding to make a valid depiction, or false by failing in some way to get it right. But maps are also performative. In fact, any map has a performative aspect in the way that it states what information is relevant. For example, a city map always has some boundary. That may be the legal city limit, the metropolitan area, the area including major landmarks, or just the most populated portion of the city. That boundary is a performative utterance, which we might paraphrase as "this part of the city is worth attending to."

An intriguing example of performative maps can be found in John Ogilby's *Britannia* atlas of 1675, the first British road atlas.[19] It set the standard for road maps to follow. One hundred strip road maps were created for the atlas, each accompanied by text advising on the map's use. An important innovation was Ogilby's scale of one inch to the mile (1/63,360). These are marked and numbered on each map, the miles further being divided into furlongs.

Ogilby had been appointed "His Majesty's Cosmographer and Geographic Printer" in 1674. His atlas has both constative and performative aspects. In his four-part television series, Terry Jones uses the strip road maps and the accompanying texts to follow Ogilby's actual route, from the English border to Aberystwyth, Wales. The series contains many surprises and mysteries that go beyond the changes one would expect from the toll of three centuries.

Some of the roads that Ogilby shows never existed at all. Others take implausible circuitous routes. Shrines and other religious sites are indicated prominently. The atlas purports to be a guide to travelers, merchants, hunters, landowners on how to traverse the roads of England and Wales. However, its unwieldy size, ample illustrations, and commentary suggest other purposes. The reliance on strip maps implies a guidance to destinations defining what is important for both the reader's conception of the kingdom and for possible military maneuvers, for instance, invasion from France.

Jones claims that one of the map's purposes was to support and rationalize the Restoration of the monarchy under Charles II after the Interregnum that Oliver Cromwell (1649–1660) led. It thus supported a Catholic renewal of the kingdom and accompanying links to France. The atlas is dedicated to Charles II.[20]

The age of colonization led to many more examples of performative maps. Maps were used not just to describe things as they are but to create new realities. They often defined territories that indigenous people occupied as "empty lands," ripe for colonization. An exception to this is the Royal Proclamation of 1763, which King George III of England issued following Great Britain's acquisition of French territory in North America. The vast region of Trans-Appalachia was delineated as an Indian Reserve (figure 5.4). It forbade all settlement west of a line drawn along the ridge of the Appalachian Mountains, preserving for a brief time some of the Indian lands.

The new reality created discontent between Britain and colonists, especially for potential settlers and land speculators. It also raised fears of "savage" violence.[21] This discontent became a contributing factor leading to the American Revolution. Amid the soaring, inspirational language of the Declaration of Independence are complaints about England's "unwarrantable jurisdiction" in affairs like this: "[King George] has excited domestic insurrections amongst us, and has endeavored to bring on the inhabitants of our frontiers, the merciless Indian Savages, whose known rule of warfare, is an undistinguished destruction of all ages, sexes and conditions."

The conflict concerning where to draw the line separating the land of colonists from that of indigenous people reveals starkly how mapmaking is not just a technical enterprise. It both reflects and redefines social reality based on who has the power. In this case, the English Crown took precedence over

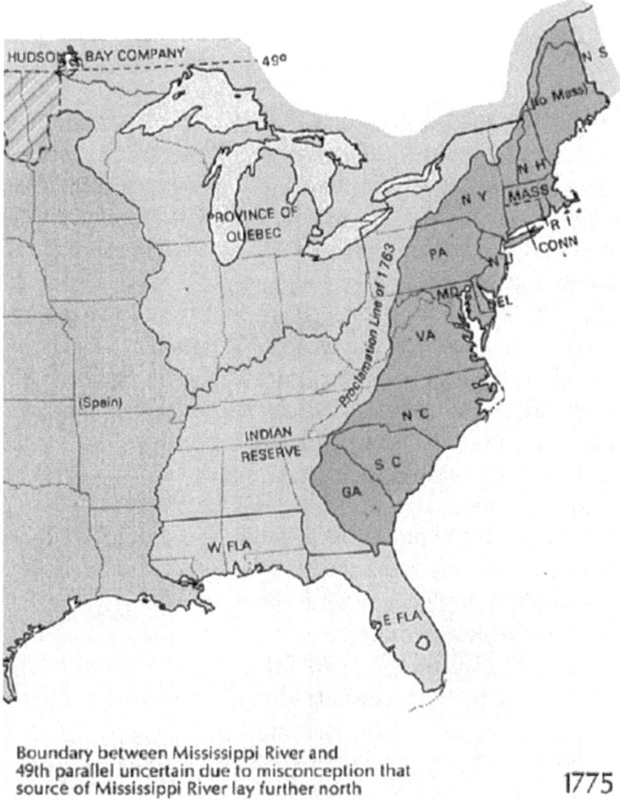

Figure 5.4. A portion of eastern North America; the 1763 proclamation line runs along the Appalachian ridge. *National Atlas of the United States, public domain*

the colonists until they rebelled and grabbed the cartographic power for themselves. It is not clear that the indigenous people were ever consulted regarding the 1763 line.

Many such striking and historically significant performative maps exist. But ordinary maps of today invariably have similar performative aspects. Some action in the world follows from any attempt to represent the way things are, not just from portrayals of what ought to be. Consider a map of geopolitical anomalies that Martin Lewis developed. It is too complex to show in full here, but even so, Lewis is well aware of what it still simplifies:

> As the title indicates, only a selection of geopolitical anomalies are indicated on the map. Many more are ignored, including the Basque nationalist movement, the Catalan nationality seeking status as a nation, the British bastion of

> Gibraltar, and the Spanish North African exclaves of Ceuta and Melilla. . . . A comprehensive map would have to be so involved as to be virtually unreadable . . . some places . . . are so geopolitically complex as to defy labels.[22]

Lewis lists nearly a hundred categories of anomalies. One, for example, is "Autonomous Country." This is the case for the Faeroe Islands and Greenland, autonomous countries within the kingdom of Denmark. Despite being under Denmark's sovereignty and, therefore, in Europe, they are not part of the European Union. As Lewis chooses anomalies to include and categorizes them in different ways, he engages in performative acts, establishing a social reality as well as describing it. Similar utterances from such official bodies as the United Nations carry even more weight, of course.

The performative function of maps is ubiquitous. Consider just municipal authorities' mundane designation of roads. Some are defined as connectors, implying higher speed limits and different regulations on signs or access. Others are central district roads with accompanying restrictions on parking and speed. They may be designated as public, subject to general regulation and maintenance; or private, allowing some local control, and imposing local responsibility for repairs.

Some roads are paper roads, meaning that the town map brings it into existence even though it winds through a thick forest and no road grader has ever touched it. Commercial street maps based on subdivision and land records similarly define streets that are public rights-of-way though impassable.

BOUNDARY OBJECTS

The research projects of Snow, Booth, and Kelley demanded individual creativity, courage, and perseverance. But none of them were individual endeavors. The same is even more true for modern mapping endeavors.

In Snow's case, Reverend Henry Whitehead was an important early critic and later supporter and collaborator, continuing Snow's work after he died. The board of health was involved as well as other municipal agencies and authorities. Various scientists promoted competing theories. The waterworks company, the church, the lithographer Charles Cheffins (who drew Snow's map), shopkeepers, ordinary neighborhood residents, and many others were involved in one way or another with the studies.

These people held diverse views about the mechanism of the spread of cholera. They also had quite different understandings of such fields as statistics and mapmaking. They undoubtedly had different reasons to be become involved and roles to play. One would not be surprised to hear of miscommunication or discord. And, yet, the projects succeeded.

Susan Leigh Star and James Griesemer point out that a common myth assumes that scientific cooperation derives from consensus, which is itself imposed by nature. But their studies of the actual work of science shows no such consensus. Instead, scientific work proceeds with internal diversity and is not held back by lack of consensus.

Star and Griesemer studied the Museum of Vertebrate Zoology in Berkeley, in particular, the research of Joseph Grinnell on speciation, migration, and the role of the environment in evolution. Grinnell's research required contributions from "university administrators, professors, research scientists, curators, amateur collectors, private sponsors and patrons, occasional field hands, government officials and members of scientific clubs."[23]

People who held these different roles had different purposes, both inherent and imposed. They had different experiences, values, and training. But they shared interest in some objects, which could transcend their social worlds. These included the following:

- species and subspecies of mammals and birds
- the terrain of the state of California
- physical factors in California's environment (e.g., temperature, rainfall, and humidity)
- the habitats of collected animal species

Thus, a map of California would be an instantly recognizable object of shared reference. "In the north" would not mean Yukon, Alaska, Siberia, or the Arctic Ocean. Instead, it would refer to the portion of the state from San Luis Obispo County to the Oregon border, or for some the area north of the San Francisco Bay area. Similarly, "cod" would not mean Atlantic codfish, but rock cod, or a species of rockfish.

The actors in this network would apply different nuances to the shared objects. For a government official, "habitat" might have less specificity than it does for an ecologist, except insofar as it overlaps with jurisdictional designations, for instance, privately owned ranch land versus federal land that allows grazing. The objects thus allow not only a general level of communication and sharing, but also divergent specificities in practice.

Star and Griesemer call such objects *boundary objects*: "Boundary objects are objects which are both plastic enough to adapt to local needs and the constraints of the several parties employing them, yet robust enough to maintain a common identity across sites. . . . The creation and management of boundary objects is a key process in developing and maintaining coherence across intersecting social worlds."[24]

EMERGENCE

By organizing data of many kinds within one map or across maps, patterns emerge that are not obvious through the examination of isolated data sets. This was the case at Hull-House for investigations of the "sweating system," slums, diet, saloons, casual labor, truancy, tuberculosis, cocaine, midwifery, infant mortality, and home reading. Data on the particular issue could be correlated with ethnicity, poverty, employment, and other factors to reveal hidden information.

Hull-House maps became *boundary objects*. Detailed representation of disease patterns, employment, and ethnicity could be presented to city officials with the aim of bringing about positive social change. The same maps, but with a different emphasis in presentation, could be shown to researchers at the University of Chicago, inspiring the nascent field of sociology. The maps could also be posted on the walls of Hull-House for the neighbors to help them understand their own situation, problems, and resources. The same map would be interpreted by different audiences based on their own backgrounds and purposes.

When the Hull-House maps were posted on the walls, they informed the neighbors, but they also helped them to see themselves as part of the same neighborhood, with common understandings and causes. The very existence of an ethnicity map reminded people of the many national origins of the immigrants as well as their common predicament. Civic intelligence[25] begins to emerge from use of these shared representations.

Maps can be incorporated into the inquiry needed for problems of public health and other social services, into public administration, to promote cultural understanding, and to solve problems in a wide variety of other fields. They are especially effective when detailed knowledge of complex phenomena is needed. Most important, they support the development of holistic understanding that engaged citizens need in a democracy.

NOTES

1. Lucy Sprague Mitchell, *Young Geographers: How They Explore the World and How They Map the World*, 4th ed. (New York: Bank Street College of Education, 1991), 82.
2. Bertram C. Bruce, *Education's Ecosystems: Learning through Life* (Lanham, MD: Rowman & Littlefield, 2020).
3. This and the following quotes are from John Snow's article, "The Cholera near Golden-Square, and at Deptford," *Medical Times and Gazette*, September 23, 1854.
4. The maps are also called dot distribution or dot density maps.

5. John Snow, "On the Communication of Cholera by Impure Thames Water," *Medical Times and Gazette*, October 7, 1854.

6. London School of Economics and Political Science Library, "Charles Booth's London: Poverty Maps and Police Notebooks," 2016, https://booth.lse.ac.uk/.

7. Residents of Hull-House, *Hull-House Maps and Papers* (New York: Arno, 1970), 3.

8. Mary Jo Deegan, "Hull-House Maps and Papers," in *Poverty in the United States: An Encyclopedia of History, Politics, and Policy*, vol. 1, edited by Gwendolyn Mink and Alice O'Connor (Santa Barbara, CA: ABC-CLIO, 2004), 386–87.

9. Cecelia Tichi, *Civic Passions: Seven Who Launched Progressive America (and What They Teach Us)* (Chapel Hill: University of North Carolina Press, 2011).

10. Residents of Hull-House, a social settlement at 335 South Halstead St., Chicago. *Hull-House Maps and Papers. A Presentation of Nationalities and Wages in a Congested District of Chicago* (New York: Thomas Y. Crowell & Co., 1895).

11. Residents of Hull-House, *Hull-House Maps and Papers*, 69.

12. Residents of Hull-House, *Hull-House Maps and Papers*, vii–viii.

13. In *Muller v. Oregon* (1908) the U.S. Supreme Court upheld Oregon state restrictions on the working hours of women as justified by the special state interest in protecting women's health. Louis D. Brandeis presented a pathbreaking brief, with only two pages of traditional legal reasoning and more than one hundred pages of sociological evidence.

14. Eleanor J. Stebner, *The Women of Hull-House: A Study in Spirituality, Vocation, and Friendship* (Albany: State University of New York Press, 1997).

15. Roger F. Tomlinson, "A Geographic Information System for Regional Planning," *Journal of Geography* 78, no. 1 (1969): 45–48.

16. David DiBiase, *The Nature of Geographic Information* (State College: Penn State's College of Earth and Mineral Sciences, n.d.), chap. 9, sec. 5, https://www.e-education.psu.edu/natureofgeoinfo/.

17. Nama Raj Budhathoki, Zorica Nedovic-Budic, and Bertram C. Bruce, "An Interdisciplinary Frame for Understanding Volunteered Geographic Information," *Geomatica* 64, no. 1 (May 7, 2010): 11–26.

18. John L. Austin, *How to Do Things with Words: The William James Lectures Delivered at Harvard University in 1955* (Oxford, UK: Clarendon Press, 1962).

19. Various facsimile editions exist, including online, https://www.fulltable.com/vts/m/map/ogilby/mna.htm.

20. *Terry Jones' Great Map Mystery*, BBC Two Wales, 2008.

21. Exact figures are difficult to obtain, but the savagery appears to be one-sided, with much more killing of Native peoples than by them. See Donald L. Fixico, "When Native Americans Were Slaughtered in the Name of 'Civilization,'" *History*, August 16, 2019, https://www.history.com/news/native-americans-genocide-united-states.

22. Martin W. Lewis, "Key to Map of Geopolitical Anomalies," *GeoCurrents* (blog), June 15, 2011, https://www.geocurrents.info/geopolitics/key-to-map-of-geopolitical-anomalies.

23. Susan Leigh Star and James R. Griesemer, "Institutional Ecology, 'Translations' and Boundary Objects: Amateurs and Professionals in Berkeley's Museum of

Vertebrate Zoology, 1907–39," *Social Studies of Science* 19, no. 3 (August 1989): 392.

24. Star and Griesemer, "Institutional Ecology, 'Translations' and Boundary Objects," 393.

25. Doug Schuler, "Cultivating Society's Civic Intelligence: Patterns for a New 'World Brain,'" *Information, Communication, and Society* 4, no. 2 (June 1, 2001): 157–81.

6

Connecting

The essential traits we associate with maps today evolved gradually over millennia. We first see cardinal directions on Babylonian clay tablet maps from five thousand years ago, for instance, but distances don't appear on maps for three thousand more years—our oldest such example is a bronze plate from China's Zhou Dynasty. Centuries more pass before we get to our oldest surviving paper map, a Greek papyrus depicting the Iberian Peninsula around the time of Christ. The first compass rose appears in the Catalan Atlas of 1375. "Choropleth" maps—those in which areas are colored differently to represent different values on some scale, like the red-and-blue maps on election night—date back only to 1826.

—Ken Jennings, *Maphead: Charting the Wide, Weird World of Geography Wonks*[1]

Ken Jennings's description of how maps have evolved, in terms of not only what they represent, for example, changing political boundaries, but also their own affordances as a medium of communication, remind us that maps are created by people to serve some purpose. For example, the compass rose became more salient as explorers undertook longer voyages across great oceans. As people began to use maps to connect ideas in different ways, the form of the maps themselves needed to change. We see that today with the advent of GPS and spatial computing.

Even one map can provide insights difficult to obtain in other ways. It can show connections and a summary view of a topic. But in general, it is with networks, when a map is connected to other maps, adjacently or in overlays, that the full power of a map emerges. It can do even more when connected to texts, images, graphs, charts, diagrams, and mathematical processes. Just as is the case for living organisms in a biological ecosystem, computers linked together, or human activity in any realm, the network leads to something greater than the sum of the parts.

For example, MassGIS (Massachusetts Bureau of Geographic Information) offers free *data layers* that can be combined to investigate all sorts of questions related to scientific inquiry, emergency response, environmental management, transportation planning, economic development, and state government operations.[2]

These layers include U.S. census data from the past three censuses, designated shellfish growing areas, bicycle trails, Massachusetts Historical Commission inventory, Boston Harbor water taxi stops, community health centers, ice rinks, bedrock lithology, Atlantic Canadian provinces, offshore town boundaries, zip codes, and much, much more. This is all information available in numerical tables, but it becomes more useful as layers that can added to a base map. The user can then see, at a glance, relationships among the data items. The whole becomes a story for engaging the world.

This chapter focuses on how maps can be created and used to promote connecting ideas and, ultimately, *civic intelligence*. Doug Schuler asks,

> If we could imagine humankind finding better responses to our myriad problems, old and new, what circumstances and resources need to be in place and what steps could be taken that would support these new responses? These preconditions and steps we can call "civic intelligence" or perhaps a "world brain."[3]

Maps are one tool to support those new responses.

RUGG'S SOCIAL STUDIES

Networks of maps exist in virtually every domain of human thought. They can, accordingly, be used to help people understand how to connect ideas from one domain to another. A fully developed example of that can be seen in the social studies curriculum that Harold Rugg and others at the Lincoln School of Teachers College at Columbia University developed in the 1920s and 1930s.

Rugg's team set out to provide an alternative to the rigid, often irrelevant education practices of the early twentieth century. The new curriculum would place individual and social needs above tradition. It would be a laboratory for working out these new ideas. Not surprisingly, they had no existing curricula or teaching materials to follow.

Developing a New Curriculum

Harold Rugg arrived at the Lincoln School and Teachers College in 1920. Although his training had been in psychology and statistics, he soon took on fully the challenge of designing the new education, especially in the area of social studies. His team faced not only the conceptual challenge of creating a new curriculum, but also of doing so without benefit of existing textbooks, teaching materials, instructional guides, or other models.

Rugg and colleagues began to develop a series of junior high school social studies pamphlets.[4] The Rugg curriculum supported both the development of individual knowledge and the preparation of a skilled workforce, thus satisfying conventional private/public needs.[5] However, it went beyond these usual goals to support what progressive educators called a reconstructionist approach, which would help in building a better society. In particular, its critical-thinking approach encouraged students to examine the values behind the facts, not simply to memorize or to apply them.

The introduction to the students emphasized how they would learn about people different from themselves, as well as answers to why the country developed as it did:

> You are beginning a year's study of interesting things about your own country and its people. You are going to learn who Americans are, where they have all come from, how they live in different parts of the United States, what kind of work they do, and many other things about the people whose life is different from yours and from others of your own town or city . . . where certain cities and mountains and important rivers are, why cities grew at some places and not at others, why railroads are where they are, the reason for locks, and canal-building, why great ports have developed.[6]

A question like, "What did the Indian think of the coming of the white man to his continent?"[7] sounds innocuous. However, it challenged a prevailing view of an empty frontier with resources available to be exploited, which belonged to European colonizers, calling instead for empathy with oppressed people. Students were thus asked to bring their private concerns and experiences into contact with more public issues and histories. This led naturally to questions about competing values, for example, unrestricted economic development versus indigenous rights.

The curriculum adopted an integrated approach. Rather than teaching history, geography, civics, economics, sociology, and so on, as separate topics, it sought a unified social studies. In so doing, it essentially created that now-familiar school subject and offered materials for teaching it. The integrated program was proposed as being more meaningful for students and less burdensome for teachers. The approach was what we would today call a social justice perspective based on students' own investigations of social problems.

Rugg thought that it was essential for teaching of social studies to match the dynamic content of American life. He thought that rebuilding the curriculum was critical, with a concentration on children's growth. Rather than providing an official version of national history, the pamphlets and later books and workbooks encouraged students to learn by examining problems and proposing potential solutions.

Rugg was not a professional historian and, accordingly, he did not always adhere to the disciplinary ideal of objectivity. He believed that a focus on unresolved social issues would make schoolwork individually and socially relevant for students. His first commitment was to progressive ideas and methods, especially helping students *do* history rather than accept the conventional wisdom about it. This commitment laid him open to attacks later on.[8]

Use of Maps

The curriculum team believed that isolated location facts were of little value, compared to understanding geographic relationships and causal chains. For example, they write, "In order to comprehend American industrial life one must understand the dominating location of Pittsburg, close to enormous stores of coal, sitting astride several of the most important lines of freight and water movement, reaching backward to a rich hinterland and forward to national and world markets."[9]

The curriculum was explicit with students about its knowledge goals (note the earlier opening quote). These goals emphasized understanding how transportation and communication technologies helped to forge a united nation:

> We want to see how they built great cities far apart from each other, and then tied them together with railroads, telephones and telegraph, Uncle Sam's postal service, and how countless newspapers carry the news of one city to another from day to day. We want to know a great deal about the important industries in the country, about our schools, about our government, so that we will know how to be good citizens and how to help America solve her problems.[10]

Accordingly, the curriculum relied heavily on maps. Figure 6.1 shows how westward expansion of the United States involved not only taking territory from England, but also acquiring territory that various European countries possessed after taking it from indigenous groups. Maps like this one supported discussions about the meaning of "ownership" of land, settlers' ideas about "empty lands," and the meaning of "progress."

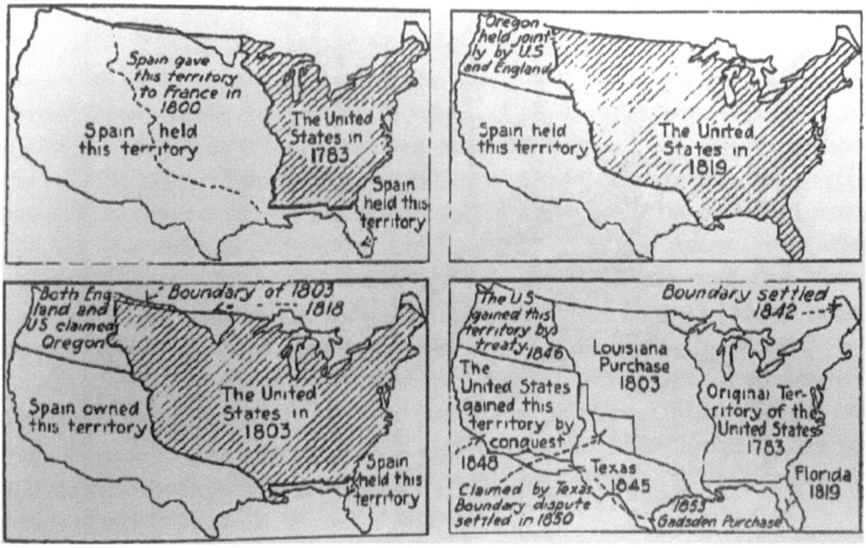

Figure 6.1. Maps of U.S. territory in 1783, 1803, 1819, and 1853. *Public domain*

The curriculum also made extensive use of other types of maps in the broad sense: timelines, flowcharts, line graphs, bar charts, tree diagrams, pie charts, pictograms, a side view of elevations for proposed canals, and more. Figure 6.2 was used in a unit called "The People Who Settled the West." It uses the width of images to represent the number of immigrants from different countries. This is a map in the broader sense of being a visual display of information. Location information is only in the labels (e.g., "Great Britain" or "The Old Immigration from Northern Europe"). Spatial relations within the map are used to convey other information, for instance, time period or numbers of immigrants.

The curriculum was Eurocentric or even U.S.-centric. This was deliberate. Rugg believed that studies in school needed to be linked to students' present-day concerns. However, despite that, and atypical for its day, the curriculum endeavored to situate U.S. social studies in a world context. It sought to help students understand the contribution of other world cultures and by implication, of the immigrants to the United States.

One interesting map shows important advances that maintained and extended scientific learning, especially during Europe's dark periods. The "torchbearers of scientific learning" includes such examples as the Indo-Arabic number system and Chinese papermaking.

Maps were used for many reasons. One was that the emphasis on children's growth meant that it was important to provide easy access to a complex body of information. Maps and pictures were supposedly easier routes to

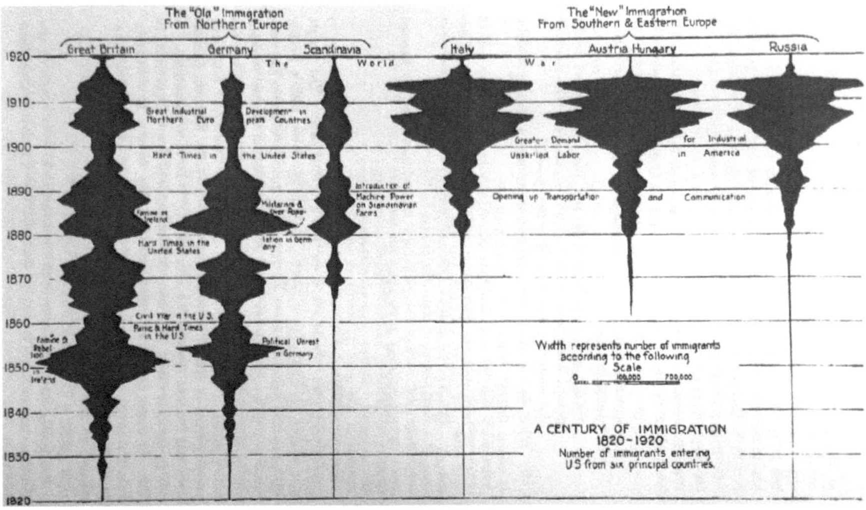

Figure 6.2. A century of immigration, 1820–1920. ***Public domain***

understanding. Also, as indicated earlier, the very content of social studies was defined by geographic relationships.

A good example of this is how the curriculum identified Ethiopia as one of the "storm centers of the world." Italy had invaded Ethiopia in 1935–1936, arousing intense fear and anger in Great Britain because it threatened the empire's main line of trade and defense, the Mediterranean–Suez–Red Sea line. This line of trade connected Britain's ports with several of her other possessions, leading all the way to India. In addition, Italy's actions endangered Britain's African colonies.

Connections among Maps

Maps in the Rugg curriculum are rarely used in isolation. More often, they invite some comparison. In some cases, they are shown as a compound of maps like the one on westward expansion of the U.S. territory, which comprises four individual maps.

"Map exercises" ask the student to construct or modify a map, again using or creating multiple maps. For example, one exercise asks the student to first outline the frontier as it stood in 1800, then color in different sections and label them according to national ownership and dates, finally to do the same for 1850. The idea is to encourage students to use map reading and mapmaking as tools for thinking. Rather than memorizing dates for significant events, they make visual and tangible the story they are learning.

Another map exercise involves collaborative learning about rivers through a series of steps. The goal is to identify principal rivers and where they flow. Exercises like this prepare the way for such discussion or writing questions as the following:

- Which was cheaper in 1800–1850, land or water transportation?
- Tell how the "lay of the land" determines where canals can be dug.
- Which city of the Atlantic coast has become largest? Why?
- See if you can find out how America compares with European countries in the way she uses water transportation.

The First Book of the Earth, the first volume in the "primary course," has an image on every other page. These include maps, images of globes, photographs, diagrams, and drawings. One shows strings connecting the Sun and planets in an attempt to explain gravity. Children were expected to use these images, not simply as motivation, but as tools for understanding. The volume contained diagrams for cell division, volcanoes, plant and animal speciation, different ways of imagining the Earth, more than a hundred images in all.

Impact and Censorship

The pamphlets sold more than 750,000 copies. Starting in 1929, they were converted into published texts, elements of a series called *Man and His Changing Society*.[11] The books and workbooks in that series were even more successful than the pamphlets had been, despite the economic depression. "Rugg and his associates truly created a unified social science course for the schools. His ideas in his books had attracted worldwide attention and imitation. He was without doubt the leading figure in social studies education in the United States."[12]

Rugg was especially concerned with material inequality:

> Like his fellow writers, he praised the nation's record of economic accomplishment. But unlike many of his predecessors, he delved into murkier topics—the pernicious influence of money in politics; the class bias of revered statesmen; and, most noticeably, the apparent failure of laissez-faire economics, which had, at the time his books rolled off the presses during the Depression's worst years, produced millions of destitute farmers and industrial laborers along with a handful of those he termed the "idle rich."[13]

By the mid-1930s, the Advertising Federation of America, the American Legion, and others began to attack the curriculum due to its prominence and its progressivism. Critics charged that the books promoted "pro-socialist ideas."

Moreover, they were deemed to be unpatriotic because of the focus on critical thinking, diverse perspectives, and a recognition of social ills and problems in American society.

A cartoon in the *American Legion Magazine* (September 1940) shows an evil teacher with a "subversive textbook." The caption says that "the 'Frontier Thinkers' are trying to sell our youth the idea that the American way of life has failed." An article refers to "Treason in the Textbooks." The *National Republic* (February 1940) says in a cartoon that the little red schoolhouse is "now being tainted red."[14]

The attacks were successful; before long, school districts began withdrawing the curriculum. However, it continued to be used. Its popularity at the ground level can be seen in the fact that the workbooks continued to sell well even after the attacks on the curriculum began. Its legacy can be seen in the continuance of unified social studies as a school subject, the use of maps in modern curricula, the reduction of a focus on the teaching of isolated geography facts, and at least some promotion of discussion and critical thinking.

Rugg had a good understanding of social studies and an ecosystem for how to teach it. He saw the value of an interdisciplinary approach that combined history, economics, political science, sociology, linguistics, and other social science disciplines using a diverse ecosystem of engaging activities and tools including text, pictures, maps, diagrams, and hands-on learning. He focused that on what seemed to him to be self-evident truths about the development of civilization, including problems related to inequality, prejudice, immigration, and industrialization.

However, he probably underestimated how his work was part of a larger ecosystem in which teaching children to think about social problems would be deemed unpatriotic. Some considered his *social reconstruction* perspective to be overly didactic; others thought it was downright subversive.

Assessment

The Rugg curriculum shows us that social studies as an integrated field of study has the potential to develop engaged citizens. Ronald Evans notes that Rugg's work "throws critical light on our collective responsibility for social justice and reminds us of our unrealized potential to create a truly democratic 'civilization of abundance, toleration, and beauty,' and to live out the meaning of our creed." Despite its ultimate demise, Rugg had a significant impact.[15]

Rugg believed that maps could support an integrated, issues-centered approach to learning. That, in turn, could help establish a more robust democracy. At the time of the implementation of his curriculum, the threats to democracy in the United States appeared to many to be located in Europe or other lands. But the response to his work showed that authoritarianism was very real at home.

Current events make this story more than a small historical observation: Did Rugg's work contribute to building a polity that would work to reinforce and expand democracy, or was it a failed attempt to do that? What aspects from that story can inform the struggle for democracy today?

DIGITAL MAPPING AS AN AVENUE FOR LEARNING

Contemporary GIS-centered projects showcase the possibilities for not only the scholarship of discovery, but also public scholarship. Students are engaging and creating for a public audience, not simply an end-of-semester term paper for an audience of one, the teacher.

GIS by Hand

In 1963, two high-school students became researchers for a now-defunct cafeteria chain as part of a summer job. They used contest entry forms to develop a primitive geographic information system (GIS) representing the source of patrons of the cafeteria. They also scouted out competing cafeterias to assess the level of current and potential business in different areas of town. The information they gathered was to help the cafeteria chain make decisions about the placement, size, and marketing of new cafeterias.

The work was somewhat tedious. They had to read the paper forms, then look up addresses in a phone book (no online search existed at that time). Then they looked up the street and address and marked it with a colored dot on a large paper map. Despite the laborious process, which would be hard to explain to a high-school student of today, they began to notice patterns through their work.

Few of the patrons listed addresses in the wealthier sections of the city. Presumably, they sought finer food than the cafeteria offered, or perhaps, what they considered to be a higher class of fellow diners. Maybe they were just less likely to fill out contest entry forms?

Also, few entries came from mostly black or Latinx areas. They were undoubtedly discouraged by signs saying, "We reserve the right to refuse service to anyone." Such signs were common in restaurants in the segregated South at that time. Some may have deemed it too expensive to eat out, even in a relatively low-cost cafeteria. Many other reasons probably existed, including at least the food selections, marketing, transportation options, and people following their friends or familiar patterns in their own life.

For two naive high-school students, the mapping opened a window into understanding aspects of their hometown, including effects of segregation, disparities in income, surnames by neighborhood, and others that we had never considered before, much less heard about in school.[16]

Even that primitive system seemed like an invasion of privacy. They couldn't anticipate how benign it was compared to social media and commerce today. They had to consider such questions as whether a lower density of dots in an area reflected lack of interest from that neighborhood or simply larger lot sizes. The process made them think about how information is collected and used. Was this summer activity even ethical, much less effective for the planning by the cafeteria chain?

Local History

A student in Lake Station, Indiana, became intrigued with Abraham Lincoln through her GIS project. This led her to visit a local historical society. She learned there that Lincoln was reported to have stayed at the town's Audubon Hotel. But her further research revealed that this was local lore that the *Chicago Tribune* perpetuated with no evidence. This experience awakened her historical consciousness.

Teachers noted other examples through this and other student projects. The original research students conducted led to learning for everyone involved:

> One student mapped the treaties signed by William Henry Harrison that essentially created three-fifths of modern Indiana, teaching us how much of our state he negotiated and acquired from the various American Indian tribes that inhabited the area. She taught us through her learning. There is probably no greater value for students than to use a medium that allows them to teach others and to share this creative work with friends on social media with a click of the mouse.[17]

Mapping Cemeteries

What sounds like a small project, or even a gloomy one, soon opened up into far-reaching explorations of history, geography, health, families, technology, mathematics, literacy, and more. At Iroquois West Middle School in Illinois, youth started with a story about a primary school's project to study cemeteries: "Learning from graveyards."[18]

The "Map Masters" soon expanded on this approach by incorporating technologies of GPS and GIS into their mapping project of the Onarga Cemetery. They made many discoveries. They also connected with cemetery mapping projects in Cass County and East St. Louis (also in Illinois). One student reported the following:

> One interesting tombstone that we found at the Onarga Cemetery was in the shape of a tree trunk. The name of the person buried there was Emory Gish. According to our research on symbolism, the tree trunk showed a life cut short. The number of broken branches might symbolize the number of deceased family members buried nearby.

Community Asset Mapping

Community asset mapping seeks sustainable community development from the inside out. It begins with discovering and sharing information about existing resources and strengths of a community rather than its deficits.[19] Creating community asset maps entails active participation of community members in raising questions, clarifying values, and investigating.

About fifteen years ago, African American youth on the north side of Champaign, Illinois, used community asset mapping to identify potential neighborhood resources. This included doing oral history and producing databases, videos, and formal presentations to record their findings. They learned that their community had a robust jazz tradition. They interviewed older musicians to hear stories about that and to hear their music. In turn, they shared their own music with the older musicians.

Other students in the community challenged the historical town-and-gown separation in a university town and, in particular, deficit views of the community. It had historically been populated by working-class African Americans, and, despite its proximity to the university, has remained socially and culturally separated from it.[20]

One community activist noted that youth felt that the university was gated, and they ought not to cross University Avenue. In response, several campus and community organizations came together to provide youth with an opportunity to cross the imaginary border and investigate resources and assets available for them locally, for example, job opportunities for teens, youth programs, and recreation centers.

Youth began to map community assets and to write their community's history. Their work required developing facility with diverse digital media together and infusing that with the drive for social change. Young people built upon their actual or latent abilities with digital photography, audio and video production and editing, internet search, GPS, databases, internet radio, and more. This meant that they could use their facility with new information tools in a way that helps connect and build community rather than leading to further isolation. Digital literacy then becomes not just another skill to acquire but a part of one's repertoire for meaningful action in the world.[21]

Their work on the project included creating flyers and distributing them, canvassing neighborhoods, interviewing residents, making videos and public presentations. These activities helped the youth mappers cultivate positive perceptions of their community, as well as develop their own social and digital literacy skills. The work that the youth mappers conducted also addressed practical community needs and challenged the deficit view of the community.

The youth mappers said that the civic project helped them learn much about their own community. Despite the challenges of canvassing (long walks, heat, dogs, rude people), the youth mappers reported that it improved their knowledge of the neighborhood geography, calling their attention to street names, signs, and the locations of community organizations. One said, "I learned about the different areas in the neighborhoods, know what streets I am on, pay attention to signs . . . I met people who have done good in the community and other youth will be interested in."

Most important, the youth mappers were surprised at the rich resources for teens their community offered. One learned about local history through an interview with a community member. In another interview, she learned about a community organization for kids with Down syndrome, and the social stigma many faced.[22]

Using technology in this way means that learners become active, critical users rather than passive recipients of technology. Their devices become part of their social practices. The online activities are networked with other aspects of life, leading to emergent properties, for instance, a growing interest in community history. This learning system is constructed by the participants, with cultural, political, and economic values embedded. Users gain agency as they adapt technical systems to achieve personal and community goals.[23]

CIVIC INTELLIGENCE

When we think about big problems like intolerance, economic justice, or climate change, should we focus on the local or the global? the short term or the long term? the social or the technical? If a goal of education is to develop critical, socially engaged citizens, then how should students learn to grapple with these issues? Questions like these recur in ways that often lead to discouragement, including the feeling that nothing can be done on any problem at any level.

Naomi Klein writes, "The very idea that we—as atomized individuals, even lots of atomized individuals—could play a significant part in stabilizing the planet's climate system, or changing the global economy, is objectively nuts." There must be a "large, organized, and focused movement." And yet, she goes on to say, local activism is critical: it is winning big fights, showing us what the future looks and feels like, and inspiring bigger examples. In short, we need both. But it can be daunting to find examples that bridge between the local and the global, the immediate need and the needed long-term change, or between social and organizational work and technological fixes. Maps are one tool to build those bridges.

Mapping as a Path to Civic Intelligence

For two decades, the coastal town of Wellfleet, Massachusetts, has hosted a free conference at the local elementary school.[24] Participants include ordinary citizens, fishermen, students from K–12 through graduate school, town officials, and staff of the National Park Service and environmental organizations. They come to discuss what they are learning about the ecosystem of the harbor.

The conferences are learning events throughout. They explore the complex interactions among humans, the environment, and other living things, including phytoplankton, striped bass, menhaden, horseshoe crabs, oysters, quahogs, seals, terrapins, molas (sunfish), phragmites, bacteria, protozoa, resident and migrating birds, as well as the land, sea, and air. For example, one recent project emphasized the rise in the ocean sunfish population attributable to increased numbers of jellyfish.

The conference is a good example of how to promote *civic intelligence*, essentially becoming more aware of the resources in our community, learning of its problems, finding ways to work together, and developing civic responsibility.[25] In any locality, civic intelligence is inseparable from the nature all around. But in this case the connection is highly evident. Every issue—transportation, affordable housing, employment, health care, fishing and shellfishing, waste management, history, and more—affects and is affected by the ability to live sustainably. The harbor and the surrounding ocean, rivers, and uplands are deeply embedded with that.

A striking feature of every presentation and poster at the conference is the use of visual displays of information. These include maps showing local geologic and geographic landmarks, tidal flows, migration patterns, seasonal variations, sediment accumulation, and human-made structures. GIS is a constant theme, using both analogue and digital techniques and tools. The most useful applications involve overlays of maps or comparisons of maps from different situations or times. Also included are flowcharts for processes, for example, one for adaptive management of tidal restoration, organization charts, and timelines for events in temporal sequences. The maps not only communicate results, but they are also necessary for the research itself.

You would find similar activities at many conferences. But this one is notable in terms of dialogue across disciplines and between professionals and citizens. Moreover, knowledge creation is integrated with daily experience and action in the world, often guided by shared maps. By engaging with nature along with fellow community members, conference attendees explore such disciplines as history, statistics, politics, commerce, geology, biology, physics, chemistry, meteorology, and oceanography. Nature revealed through maps is the curriculum guide.

Maps to Serve Society

One project that has made progress on Klein's agenda is Kathmandu Living Labs (KLL) in Nepal.[26] The project engages young people in making maps to serve societal needs. They use the online platform OpenStreetMaps.

KLL uses online mapping to address a wide variety of social challenges. Their work can be seen most dramatically in the timeline of their response to the 7.8 magnitude earthquake that devastated Nepal on April 25, 2015, causing widespread damage across fourteen districts.[27] A portion of that timeline is summarized below.

After the quake, KLL sought help from the international OpenStreetMap community. Operating from a parking lot because its office was damaged in the quake, KLL established a situation room as the headquarters for its earthquake response. It used QuakeMap to bridge the information gap between the quake victims and relief agencies.

Within forty-eight hours, more than fifteen hundred people had used aerial imagery to remotely map the affected area in OpenStreetMap. KLL coordinated with mappers throughout the world using chat rooms. Individuals, volunteer groups, and humanitarian organizations requested data as well as printable maps for relief operations.

By April 29, KLL had established contact with the Nepal army through its GIS division. KLL coordinated volunteer mappers to map camps of displaced people. The next day they held a training session for doctors to teach them how to use OpenStreetMap and QuakeMap to determine where medical help was needed. By April 30, thirty-three hundred mappers had assisted in quake-related mapping (figure 6.3).

On May 1, KLL introduced a repository of printable maps based on the data that volunteers had added to OpenStreetMap. They began working to document the condition of cultural heritage sites in the Kathmandu Valley. They also developed a mobile data collection app and provided training for volunteers on how to report the conditions of sites through that.

QuakeMap continued to be a vital tool in the earthquake response with fifteen hundred reports about the needs of earthquake victims and relief efforts. On May 24, a landslide blocked the Kali Gandaki River. It created a

Figure 6.3. Middle school student in Nepal mapping his community. ***Kathmandu Living Labs, used with permission***

temporary dam that led to massive flooding upstream, showing the need for continued mapping.

On July 13, the Department of Education and the World Bank asked KLL to support an assessment of damage to schools. With their partners, they organized a workshop on using drones for humanitarian work. The high-quality aerial imagery that the drones collect can be used for assessing damage and planning reconstruction.

Through KLL, a farmer in a remote Nepalese village can integrate his local knowledge of a field or stream with that of an engineer at MIT. People in Nepal, despite their own challenges, were among the largest contributors to relief efforts following Typhoon Haiyan in the Philippines. Detailed information about a school can be used to influence government policies and actions. This connecting across levels or communities is something many people talk about but few have shown how to achieve.

For example, KLL has made maps to promote earthquake resilience, a very real challenge in Nepal, especially since the 2015 earthquake. It is important to know the spatial relation of such crucial facilities as health centers and schools to fault lines. It is also vital to know which roads are passable, where public safety facilities are located, and more. KLL gathers this information through satellite images, direct observation and investigation, surveys, and other techniques.

Good maps are useful for all sorts of purposes, including the much-needed economic development in Nepal. But the project offers much more than just producing maps. It can answer specific questions, for example, why some residents were not taking advantage of the government's rebuilding program after the earthquake. Some would not apply for rebuilding grants at all; others would do only phase one of three.

The answer, which came from surveys and map analysis, was that the government required earthquake resilient construction, meaning essentially concrete with rebar. But in remote villages it was difficult to get the necessary materials because of damaged roads. Villagers lacked crucial equipment and expertise. Many young people, possible workers, had left to work in Malaysia or the Gulf States. They would send remittances but were not present in person for rebuilding. The result of all of this was that villagers could not comply with the grant requirements.

Digital Internship and Leadership

Beyond general maps and specific questions, KLL demonstrates an emergent property of mapmaking. When young people learn how to make digital maps, they learn technical skills, for example, those required for GPS and GIS. They also learn collaboration, presentation, and leadership skills. The Digital Internship and Leadership (DIAL) program is designed to fill data gaps in rural Nepal. It integrates targeted mapping, a virtual internship strategy, and youth leadership development.

DIAL interns are recent graduates and undergraduate students from various academic backgrounds. They participate in a four-day in-person orientation followed by two months' internship focusing on mapping. They then return home to various areas of rural Nepal, for instance, in Kanchanpur, Mugu, or Jajarkot districts, the flatlands of western Nepal, and flood-affected areas of eastern Nepal, and continue mapping from there.

One intern in the DIAL program recognized new ways of seeing: "Something I noticed was that I started looking around and noticing places I would probably have ignored; mapping made me aware of my surroundings, and I would discover facts and details I might not have discovered otherwise." Another intern saw the value of maps for reading society: "Reading about the aspects of society that a map can reflect and realizing it practically by mapping was illuminating; I got to glimpse the socioeconomic conditions of people of different parts of Nepal." Another intern appreciated field data collection as offering experience with "how to deal with people in communities" and how to access the local funds of knowledge for public benefit.[28]

This work is ongoing, but it already demonstrates the emergence of *map consciousness*, the ability to use maps to attain an overview of a territory and a situation. Just as maps have facilitated geographic discoveries, scientific explorations, urban planning, public health, and work in many other fields, they can become tools for young people to learn about their community and society, to understand both the problems and the resources, and to become socially responsible, engaged citizens.

CONNECTING IDEAS

Maps reveal connections; they tell us how such-and-such place (or idea) is related to another. An individual map does this: It can show the direction and distance of another location. It can show river valleys and mountain ranges, cities and harbors, and much more. That mission to show connections can be partially achieved if the map stands on its own, unrelated to other maps, because a good map models its subject matter.

However, the connecting function is circumscribed if the map does not relate to other maps and other representations of knowledge. Its fulfillment comes when the map is in rich dialogue with other maps, indexes, textual descriptions, images, activity guides, and devices linking it the external world. As part of a network, one in which the whole is greater than the sum of the parts, the map shows the context for ideas. As the network develops, especially as it invites collaboration, it connects the user to the larger human community.

When a map user becomes an active map reader, she interprets it and thereby adds to its nexus of connections. When she modifies the map by adding layers, pinning landmarks, coloring sections, editing features, adjusting scale, and so on,[29] she can make connections more evident or suggest entirely new ones. When she creates a map de novo, she does all of this and, especially, expresses the connection to her own life and experience.

NOTES

1. Ken Jennings, *Maphead: Charting the Wide, Weird World of Geography Wonks* (New York: Simon & Schuster, 2011), 8. One reason for these long gaps is that many early maps have not survived.

2. Massachusetts Bureau of Geographic Information, https://www.mass.gov/orgs/massgis-bureau-of-geographic-information

3. Civic intelligence is knowledge and skills devoted to addressing public or civic issues. It applies to individuals, but it can be extended to organizations, institutions, or societies. Greater civic intelligence should increase political awareness and engagement. See Doug Schuler, "Cultivating Society's Civic Intelligence: Patterns for a New 'World Brain,'" *Information, Communication, and Society* 4, no. 2 (June 1, 2001): 164.

4. Ronald W. Evans, *This Happened in America: Harold Rugg and the Censure of Social Studies* (Charlotte, NC: Information Age, 2007).

5. Harold Ordway Rugg, John Hockett, and Emma Schweppe, *Objective Studies in Map Location* (New York: Lincoln School of Teachers College, 1925).

6. Harold Rugg, Earle Rugg, and Emma Schweppe, *The Social Science Pamphlets* (Charleston, SC: BiblioLife, 2010), iii.

7. Rugg, Rugg, and Schweppe, *Social Science Pamphlets*, 34.

8. Joseph Moreau, *Schoolbook Nation: Conflicts over American History Textbooks from the Civil War to the Present* (Ann Arbor: University of Michigan Press, 2003), 220.

9. Rugg, Hockett, and Schweppe, *Objective Studies in Map Location*, 33.

10. Rugg, Rugg, and Schweppe, *Social Science Pamphlets*, iii.

11. Elmer A. Winters, "Man and His Changing Society: The Textbooks of Harold Rugg," *History of Education Quarterly* 7, no. 4 (December 1, 1967): 493–514.

12. Winters, "Man and His Changing Society," 510.

13. Moreau, *Schoolbook Nation*, 221.

14. Evans, *This Happened in America.*

15. Evans, *This Happened in America*, xiii.

16. Author's personal experience.

17. Christopher J. Young and Joseph Ferrandino, "The Old Is New Again: Digital Mapping as an Avenue for Student Learning," *EduCause Review*, October 8, 2018.

18. Caroline S. Donnan, "Following Our Forebears' Footsteps: From Expedition to Understanding," September 29, 2000, 1–6.

19. This approach has been codified especially through the widely used asset-based community development (ABCD) model. See John P. Kretzmann and John L. McKnight, *A Guide to Capacity Inventories: Mobilizing the Community Skills of Local Residents* (Chicago: ACTA Publications, 1997).

20. This occurred through the *Engaging and Empowering Youth (E2Y)* project. See Chaebong Nam, "Crossing through the Invisible Gate, Mapping Our Neighborhood: The Engaging and Empowering Project (E2Y)," *Journal of Community Engagement and Scholarship*, May 25, 2013.

21. This example is adapted from Bertram C. Bruce, *Education's Ecosystems: Learning through Life* (Lanham, MD: Rowman & Littlefield, 2020).

22. Nam, "Crossing through the Invisible Gate," 7.

23. Martin Wolske, "Demystifying Technology: Community Inquiry for Social Change and Transformative Action," 1–13 (Prato, Italy: Monash Centre, 2014).

24. See Wellfleet Bay Wildlife Sanctuary, https://www.massaudubon.org.

25. Doug Schuler, "Cultivating Society's Civic Intelligence: Patterns for a New 'World Brain,'" *Information, Communication, and Society* 4, no. 2 (June 1, 2001): 157–81.

26. Naomi Bloch, "GIS and the Global Community: Humanitarian Mapping," *Geothink: Canadian Geospatial Open Data Research Partnership* (blog), November 18, 2015, http://geothink.ca/gis-the-global-community-humanitarian-mapping.

27. Kathmandu Living Labs: Six-Months of Earthquake Response, http://kathmandulivinglabs.github.io/timeline-sixmonths.

28. Kshitiz Khanal, Nama Raj Budhathoki, and Nancy Erbstein, "Filling OpenStreetMap Data Gaps in Rural Nepal: A Digital Youth Internship and Leadership Programme," *Open Geospatial Data, Software and Standards* 4, no. 1 (December 2019): 7.

29. Lisa Berry, "Six Easy Ways to Improve Your Maps Mapping," *ArcGIS Blog* (blog), April 12, 2016, https://www.esri.com/arcgis-blog/products/mapping/mapping/6-easy-ways-to-improve-your-maps/.

7

Entering the Map

> There was once a boy named Milo who didn't know what to do with himself—not just sometimes, but always . . . "I can't see the point in learning to solve useless problems, or subtracting turnips from turnips, or knowing where Ethiopia is or how to spell February." And, since no one bothered to explain otherwise, he regarded the process of seeking knowledge as the greatest waste of time of all.
>
> —Norton Juster, *The Phantom Tollbooth*[1]

Milo thinks that learning things is a huge waste of time, but he does not know what else to do with himself. One day he receives a surprise package containing a genuine turnpike tollbooth, precautionary signs, assorted coins for use in paying tolls, a book of rules and traffic regulations, and "a beautiful map, in many colors, showing principal roads, rivers and seas, towns and cities, mountains and valleys, intersections and detours, and sites of outstanding interest both beautiful and historic."[2]

There is also a car. Milo walks across the room and dusts off the car. He hops in and drives up to the tollbooth. As he deposits his coin and rolls past, he remarks wistfully, "I do hope this is an interesting game, otherwise the afternoon will be so terribly dull."[3] So begins the adventure of his life.

The map for the phantom tollbooth imagines a realm to be explored, enjoyed, and challenged by. Like a dollhouse, a model train set, or just a simple set of blocks, it presents an alternate world. On the map are such areas as Dictionopolis, Mountains of Ignorance, and Foothills of Confusion. It may be mysterious, even scary at first, but it is ultimately manipulable and comprehensible.

For Milo, the map is not an object of study or a reference to confirm details of his surroundings, but a portal into a parallel reality. He does not read or refer to the map; he enters it and becomes one with it. His map is, thus, very personal. The story told through traversing the map is unique to his own life, his prejudices, his knowledge. As we explore the map with him, we learn about both Milo and the world.

Milo does not know where he wants to go, so his "beautiful map" cannot be the answer to an unasked question regarding how to get somewhere. Instead, it is a talisman offering adventure, knowledge, wisdom, power, and other benefits. It invites Milo to explore. As he does, he makes and remakes the map through his own questions and experiences.

We experience a similar process when we explore a new region, or even a familiar region in a new way. Our map could be a blank sheet of paper, which we embellish with the learnings from our journey. Like Milo, we enter that map and create it as we use it. It then becomes a tool to communicate our new, situated knowledge.

COMMUNICATING WITH MAPS

In their excellent guidebook *Mapping*,[4] Gary Benenson and James Neujahr point out that maps are, above all, a means of communication. We can convey ideas in many ways. We can use written or spoken words; numbers; two-dimensional graphics; three-dimensional models; other sound-based means of communication and expression like music, gestures, and body language; three-dimensional sculptures and models; and, often neglected, maps of many kinds. To be the most effective communicators, we need to learn to both read and write maps, graphs, diagrams, Web pages, and other graphic representations.

This is not to deny the value of ineffable experience as a source of learning. We may encounter some phenomenon, interact with it, and find ourselves changed by that process. That can happen without anyone else knowing about it or sharing their own similar experience.

However, when we use language (broadly conceived) to express our reactions or to hear from others, learning moves to a higher plane. This is why the first two R's, reading and writing, are essential for learning, and why 'rithmetic, the third R, is increasingly seen as a mode of communication.

We may think of maps as objects to consult online, or perhaps buy for a collection. But we can make our own maps. In fact, the process of constructing concept maps, informal place maps, timelines, diagrams, and such is an integral part of thinking in almost any domain. A visual depiction of an idea allows us to articulate it more clearly, interrogate it, and share it with others.

Mapping does not fit neatly in the traditional curriculum, although it overlaps with every other area: "Nearly every topic becomes easier to understand when graphic devices are used. Imagine trying to understand history without timelines, music without scores, or geography without maps."[5] This chapter explores a vital aspect of learning: not only maps, but *mapping*, the process of telling a story through a map you create.

LYING WITH MAPS

Despite the power of maps to communicate, their use is never a straightforward depiction and transmission of facts about the world. In *How to Lie with Maps*, Mark Monmonnier makes a convincing case that maps do not just occasionally misrepresent reality; they must do so. Moreover, their usefulness is directly tied to the kinds and degrees of distortions they make. Maps are rhetorical devices, just like written texts. No text can be constructed as complete or unbiased; each one has a context, a purpose, and an audience that shapes it. Similarly, no map reproduces reality, nor should it attempt to do so:

> Not only is it easy to lie with Maps, it's essential. To portray meaningful relationships for a complex, three-dimensional world on a flat sheet of paper or a video screen, a map must distort reality. As a scale model, the map must use symbols that almost always are proportionately much bigger or thicker than the features they represent. To avoid hiding critical information in a fog of detail the map must offer selective, incomplete view of reality. There is no escape from the cartographic paradox: to present a useful and truthful picture, an accurate map must tell white lies.[6]

The cartographic paradox that Monmonnier describes may seem to be a problem, but it is actually the driving force for how we can learn by making maps. The cartographer must consider purpose and audience in making a map, making choices about what to say, what to highlight, and what to leave out.

Maps can be fictional, too, whether it is Milo's map for his journey across the "Sea of Knowledge" or those in *Treasure Island*, *Dune*, *Flatland*, *Lord of the Rings*, *The Princess Bride*, *Winnie-the-Pooh*, *Twenty Thousand Leagues under the Sea*, *Gulliver's Travels*, *Atiye (The Gift)*, and many other tales. They are a staple in detective stories, and in war stories, in fantasy and science fiction, making the supernatural world more comprehensible and believable. They arise in scenario planning and imaginary play.

Learning about those choices makes any user of a map a more critical reader and thinker. It makes him more reflective about his own processes of thinking, as well as his literacy applied to texts, maps, and other representations.

YOUNG GEOGRAPHERS

The subtitle of Lucy Sprague Mitchell's book *Young Geographers* (1934) is "How They Explore the World and How They Map the World." That expresses two key ideas that stand out even today: The first emphasizes the active process of exploration in learning, not simply learning as the reception of the ideas of others. The second identifies maps as a means of reflecting upon and articulating the results of explorations.

Mapping to Learn

Mitchell viewed mapping much as others later characterized writing: any language is a tool for learning. It is "a way for living beings to take in and enjoy the world: a means of apprehending, ordering, and heightening relationships in a way which brings on the experience of new ones. Indeed, through language (as through any art) we not only savor experience: we create it."[7]

Mitchell's great insight was that children are natural explorers, drawing upon their sense and imagination. The here and now was, for them, a stranger-than-fiction present.[8] Reading about their own world, or better yet, re-creating it through language and maps, was the path to learning.

Several decades later, the writing-to-learn movement that Janet Emig, Toby Fulwiller, Lee Odell, Arthur Applebee, and many others exemplified extended Mitchell's insights. A key event was James Britton et al.'s study of the development of writing by 11- to 18-year-olds. They identified three functions of writing: transactional writing to communicate information, poetic writing to create beautiful objects, and expressive writing to explore and reflect upon ideas.[9]

The third category, expressive writing, is crucial for learning at every developmental stage, because language is a major way in which we organize experience. A practical implication is that cross-curricular writing programs can enhance student learning. The writing-to-learn movement thus wrenched writing from its usual roles as a means of assessing knowledge transfer or as stylistic performance and saw it instead as a tool for thinking.

Mitchell would have welcomed the writing-to-learn movement. She had seen much earlier that maps were a language for expression and learning. This is why she emphasized the use of *tool maps*, which children create and manipulate, as well as *reference maps*, which provide information that one can study and acquire information from. This distinction is comparable to that between writing texts and reading texts as means for learning.

Learning Geography

In contrast with some modern writers who lament the supposed loss of geographic content knowledge, Mitchell did not position geographic knowledge or map skills as the end point for learning. Much more basic was the idea of exploring the world, a complex process in which maps could be an invaluable aid. The young geographer was an explorer first, one who learned about maps and their uses as part of that process of exploration.

Children would extend their learning about and through maps by going out into the city or countryside. They would then use maps to record and make sense of their findings. Figure 7.1 shows children playing city work activities after trips to the Manhattan waterfront. They use blue oilcloth to represent the river and block towers for the buildings. They construct bridges and ships. Through this tool map, they articulate and reimagine what they have learned and further explore its meaning.

Mitchell saw exploration as a lifelong process. The school could enhance that process, but it would not initiate it or bring it to a final resolution. She writes,

Figure 7.1. Children making and exploring a map at Public School 41 in New York, 1920s. ***Courtesy of Record Group 4—Institutional Advancement, Bank Street College Archives, New York***

> The study of the environment begins unofficially at birth and could continue throughout life. It is obvious that the environment can be a field for exploration from the time when a baby has not discovered that his toe belongs to his body, to the time when he discovers that the social thinking of his family or his nation is one of many kinds of "clan thinking" produced by different differing cultures in differing environments.[10]

Figure 7.2 shows how mapping mediates among reading, discussions, art, flat maps, and 3-D objects. The children have constructed and are now painting a model they made of the colonial settlement at Jamestown, using wood, papier-mâché, glue, and other materials.

Projecting on a Plane

Mitchell believed that such concepts as *projection* were not just important vocabulary items, or keys to unlocking how maps worked. Crucially, they were opportunities for problem solving and learning. She writes:

> A map is a symbol, a representation of the earth surface, or some portion of it, indicating the relative size and shape or distribution of given surface features. We are here concerned with the "base map" not with the wide variety of data

Figure 7.2. Ten- and eleven-year-olds painting their model of the colonial settlement at Jamestown, Virginia. Wood, papier-mâché, cardboard, glue, and paint were the primary materials. ***Courtesy of Record Group 4—Institutional Advancement, Bank Street College Archives, New York***

> which may be recorded upon it. The problem of the "base map" when it is to be flat (not a globe) is the problem of representing a few spherical surface (the Earth) or a part of it upon a flat surface—describing a three-dimensional object in a two-dimensional convention. The technique by which this is accomplished is called a projection—the sphere is projected upon a plane.[11]

Some curricula might ignore the problem and the various techniques for addressing it, on the grounds that it is too arcane or difficult to understand, certainly beyond the grasp of ten-year-olds. Others might mention it and briefly discuss various projections with their pros and cons. Mitchell, in contrast, puts the problem to the children "concretely."

They are given a globe, or even a large ball or orange. They are asked how one could describe a particular location, perhaps a lonely island in mid-ocean. Some suggest measuring from the North Pole. That leads to the question "where is the North Pole?" That, in turn, leads to discussions of compasses, the magnetic pole, the North Star, and the spinning of the Earth on its axis.

From there the children discuss how

> North and South poles are absolute points for earth dwellers; that is, they are north and south in relation to all other points on the earth's surface. [This is not true for any other points on the sphere.] Discovery of these relations often leads to a discussion of the way the sun's rays strike the earth surface, leading in turn to seasons, long days and short nights or vice versa, and to the height of the sun above the horizon in different parts of the earth surface of different seasons. . . . With older children such words as zones, equinox, and orbit may acquire meeting. The Equator now emerges not merely as a line halfway between the Poles, but as a line over which the sun at noon stands directly overhead at the time of the year when the earth is exactly equal to the night (equinox).[12]

Mitchell goes on to describe the many ideas that open up when the children are allowed to ask their own questions and pursue their own answers. Maps stimulate, encourage, and guide the inquiry. Even if not every class proceeds as described, the case for maps to learn is convincing.

At some point, children are encouraged to create their own projections, again, well before learning the details or names of standard projections, for example, Mercator. A common approach is to propose (and create on paper or with an orange peel) a daisy-like representation with the North Pole at the center. It becomes evident that the display "splits." It is impossible to maintain both continuity and spatial relations without spreading out the display. Children learn by doing why polar maps in textbooks are never drawn beyond the equator: A map centered on the North Pole would otherwise result in an enormous circle for the South Pole.

Once children have been invited to ask their own questions about maps and solve long-standing problems in cartography, they are better able to appreciate, understand, and remember maps from the literature, for instance, the Goode equal area projection[13] or the two-hemisphere projection. Through construction they enhance experience by creating the artifacts that define it.

The most comprehensive and sustained example of a child articulating his experience through maps can be found in the novel (and movie) about T. S. Spivet, a ten-year-old cartographer. As he says, his notebooks contain maps covering almost every action performed on the family farm, including

> diverting the irrigation ditches, mending the fences, rounding up the cattle, branding, shoeing, haying, vaccinating, and castrating (!), breaking the cayuses, slaughtering the hens&pigs&rabbits, plucking the hackberries, trimming back the bracken, p-icking and shucking the sweet corn, mowing, sweeping, cleaning the tack, coiling all those lariats, oiling the old Silver King tractor, and punching the goats' heads out of the fences to keep 'em from the coyotes.[14]

The final map in the book is Spivet's Map of Loneliness.[15] It is an outline drawing of a street in Chicago, with one person walking, and dashes indicating her path. Below, a pie chart summarizes his observations of group sizes. Another indicates how many of those walking alone were using headphones or mobile devices. There is even a drawing of a headphone set. As we follow Spivet on his fantastic journey, we see how maps can be used to express every aspect of our experiences and perceptions.

PUTTING ONESELF ON A MAP

In *Cosmic View: The Universe in 40 Jumps*, Dutch educator Kees Boeke combines writing and graphics to explore many levels of size and structure, from the astronomically vast to the atomically tiny.[16] The book begins with a photograph of a Dutch girl sitting outside a school, holding a cat. It is drawn on a scale of 1 to 10, meaning that a centimeter on the drawing is in reality ten centimeters. Each successive page backs up from the original photo, with graphics that include ten times more space around the girl. It then narrows in on the original picture, with graphics that show ever smaller areas until the nucleus of a sodium atom is reached.

This idea has been used in many other books, films, and museum exhibits. In *Powers of Ten*, the reader is taken on a spectacular journey through forty-two consecutive scenes, from the dimension of one billion light years to the realm of the atom. Each scene is at a different "power of ten" level of magnification. Illustrations, essentially maps of each level, depict what we can perceive at each progressively smaller level of magnitude.[17]

Sometimes, the emphasis is on levels with meaning other than simply a jump in order of magnitude. A popular children's book, *Me on the Map*,[18] follows a young girl as she finds her place on maps of her room, her house, her street, her town, her state, her country, and her world. Returning to the local, she says, "Just think. . . . In rooms, in houses, in streets, in towns, in countries all over the world, everybody has their own special place on the map. Just like me. Just like me on the map."

In addition to challenging us to think about scale and the enormous range of phenomena in the universe, these works also serve to bring us into the map. We begin to think of a galaxy not only as an enormous object "out there," but also as a way of defining the space we inhabit, because we each have a location in the Milky Way, as well as on our local street.

This positioning is a theme throughout Lucy Sprague Mitchell's work from a century ago. It continues in the work of Bank Street College. Future teachers are asked to map the space they live and work in or their route to the college. In the geography class that Abby Kerlin and Ellen McCrum co-teach, students walk around the block and then make a map showing what they saw and heard (see one example in figure 7.3). One might focus on types of buildings and street names. Another might emphasize vegetation. Some report on the alleyway that cuts through the block, which others ignore. This experience helps them to see how a map is a form of writing in which the author's perspective and purpose are as important as the details on the ground.

Figure 7.3. Student map of the city block around Bank Street College. ***Sketch by Ian Dudley for geography class in 2017, taught by Ellen McCrum and Abby Kerlin, Bank Street College of Education, used with permission***

DEEP MAPPING

A recent study of aging examined the implications of *awe walks*.[19] Participants were asked to take walks alone in any outdoor setting (urban or rural) at a moderate pace. They were asked to refrain from using their phones except to take photographs. In one group, participants were also told that with the right outlook, awe can be found almost anywhere, but most likely where there was physical vastness or novelty. They were asked to tap into their sense of wonder and to go somewhere new each week, if possible.

The simple intervention led participants to attend to the details of the world and enhanced their sense of wonder. They had greater feelings of social connection, kindness, and humility. Throughout time, even their selfie photos changed. There were bigger smiles and a shrinking self; they filled proportionately less of their photographs with their own faces and more with the background scenery.

Participants had no problem finding the route for their walk or for getting back home, not needing a map for that. But they were making a kind of map, which showed their path in a deeper way. As they learned to observe the place outside themselves, their experience oddly became more personal, more attentive to their own lived experience.

These walkers saw things they had not noticed before and thought more about their own role in the space. This was the beginning of a *deep map*, one that embraces the *lore of place*. Deep mapping is a new type of cartography, one that includes all of the local lived experience, reaching into its deep past.[20]

Tim Robinson was a leading exemplar of deep mapping. Most notably, based on his feeling that traditional maps concealed more than they revealed, he set out to deep map Inis Mór (Inishmore), the largest of the Aran Islands in Galway Bay in Ireland, and the largest island off the Irish coast with no bridge or causeway to the mainland. For the deep mapping task, Robinson had to immerse himself in the local culture, learn Irish, and study the island's literary, scientific, and historical contexts.

Robinson's deep map is a process of discovery, but it also has representation in field notes and detailed traditional maps, now available online.[21] For example, there are handwritten field cards for the townland of Eoghanacht. Each card describes a feature—geologic, historical, or ecclesiastical. Each feature is assigned a number corresponding to a marking on one of his copies of the six-inch Ordnance Survey map. A more detailed narrative and description is in two volumes. *Stones of Aran: Pilgrimage* (1986) focuses on the coast and *Stones of Aran: Labyrinth* (1995) examines the interior.

For an exhibition a decade later, Robinson enlarged his map of Aran.

The result was a 22-foot-long map of the three islands, at a scale that invites one to step from island to island like a giant of the Fir Bolg, the islands' mythical inhabitants in larger times. This floor map was glued down in a rough old school courtyard, and was accompanied by a notice embodying a discrete challenge to the public:

> The original Aran map was printed in an edition of several thousands, and when one copy was worn out another could be obtained. But the present vinyl addition is unique, and what will become of it remains to be seen. You are invited to walk on it, to dance on it, to write on it, to treat it as you see fit.[22]

Exhibition goers would walk on the map, occasionally adding "a few skateboard marks and some rather touchingly romantic graffiti." Thus, this deep map of the islands was itself further mapped by those visitors. Later "The Distressed Map of Aran" became yet another exhibit. The various manifestations of Aran—online repositories, books, maps, museum exhibitions, and more—all manifest Robinson's view of mapping as participation. The map is not a fixed or canonical representation of experience, but it is an aid to richer and fuller engagement, which itself changes through that engagement.

Deep mapping is a practice, a form of place making, rather than a medium, for example, an atlas or a globe. In traditional cartography, the place being mapped is conceived as a stable, bounded, or fixed object; the map ignores its transient features, for example, the effect of the seasons, and emphasizes its relatively invariant ones, for instance, place names.

In contrast, deep mapping implies a continuous process of delving deeper into the art, music, sounds, smells, legends, language, and history of a place. A one-time event may be salient for understanding the meaning of the place. An ancient, no longer used, place name may be more important than the current designation, especially if it is simply some regularized name for the convenience of drivers on a motorway. It follows that the practice of deep mapping is inevitably linked with *deep travel*, most typically through walking, in which one can observe and interact in detail, learning not only about the place per se, but also one's own role within it.

Deep mapping is done in a wide variety of ways, and those are changing through its practice. Nevertheless, several key features stand out. Deep mapping is a personal process, in which the cartographer's interest, background, time and energy, and more all play a role. It is historically grounded. It emphasizes what has been in a place as well as what is evident from a surface examination today. It is multimodal, drawing on experiences of light and sound, music, myths and legends, nature, history, and philosophical values.

LABORATORIES AND STORIES

Robinson writes in a later book, "I am aware of the selectivity of my written response to living in Connemara. I concentrate on just three factors whose influences permeate the structures of everyday life here: the sound of the past, the language we breathe, and our frontage onto the natural world."[23] Deep maps invite dialogue among the variety of perspectives investing a place, past and present. The cartographer expresses her relation to place through experiencing it. Detailed inquiry about its multilayered aspects enables constructing that place in a new way.

When we view Robinson's maps, or those of Mitchell's young geographers, we are reminded that maps of many kinds offer rich media for reading the stories that others tell. The deeper a map becomes, the more it allows us to experience a strange world vicariously. The complement of this is that the media of maps can be used to write stories as well as to read them. As we articulate our own experience, we may find that our memory of and our attitude toward that experience changes. The map is not just a communication channel, it is the site for sense making.

The perspective of writing is especially important when we follow Lucy Sprague Mitchell's call to think of geography as a laboratory pursuit. After all, inquiry is at its core another word for story. Just as any scientist would do, students seek source material in the world around them or in texts by other inquirers. They then invent tools to study the relationships inherent in these data. Their maps become stories about their explorations and their findings. They become skilled cartographers, but even more, they learn to enter the world with confidence and critical awareness.

NOTES

1. Norton Juster, *The Phantom Tollbooth* (New York: Random House, 1988), 9.
2. Juster, *Phantom Tollbooth*, 14.
3. Juster, *Phantom Tollbooth*, 15.
4. Gary Benenson and James Neujahr, *Mapping* (Portsmouth, NH: Heinemann, 2002), 20.
5. Benenson and Neujahr, *Mapping*, 20.
6. Mark Monmonnier, *How to Lie with Maps*, 3rd ed. (Chicago: University of Chicago Press, 2018), 1.
7. Mary Phelps and Margaret Wise Brown, "Lucy Sprague Mitchell," *Horn Book Magazine*, June 1937. Reprinted in *Educate*, https://educate.bankstreet.edu/books/18.
8. Mitchell's *Here and Now Story Books* showed that ordinary life could be intriguing and educational for children.

9. James Britton et al., *The Development of Writing Abilities 11–18* (London: Macmillan, 1975).

10. Lucy Sprague Mitchell, *Young Geographers: How They Explore the World and How They Map the World*, 4th ed. (New York: Bank Street College of Education, 1991), 9.

11. Mitchell, *Young Geographers*, 34–35.

12. Mitchell, *Young Geographers*, 35.

13. J. Paul Goode's *homolosine* projection, https://pro.arcgis.com/en/pro-app/help/mapping/properties/goode-homolosine.htm.

14. Reif Larsen, *The Selected Works of T. S. Spivet* (New York: Penguin, 2009), 32.

15. Larsen, *Selected Works of T. S. Spivet*, 374.

16. Kees Boeke, *Cosmic View: The Universe in Forty Jumps* (New York: John Day, 1957).

17. Philip Morrison and Phylis Morrison, *Powers of Ten: About the Relative Size of Things in the Universe and the Effect of Adding Another Zero* (New York: Freeman, 1982).

18. Joan Sweeney, *Me on the Map* (New York: Scholastic, 1996). See also Loreen Leedy, *Mapping Penny's World* (Square Fish, 2000), which takes the reader via maps from Penny's bedroom to the entire world.

19. Virginia E. Sturm, Samir Datta, Ashlin R. K. Roy, Isabel J. Sible, Eena L. Kosik, Christina R. Veziris, Tiffany E. Chow, et al., "Big Smile, Small Self: Awe Walks Promote Prosocial Positive Emotions in Older Adults," *Emotion*, September 21, 2020.

20. Deep mapping is most notable in the artistic and philosophical practice of Tim Robinson. See Jos Smith, "Anticipating Deep Mapping: Tracing the Spatial Practice of Tim Robinson," *Humanities* 4, no. 3 (July 21, 2015): 283–303; Eamonn Wall, "Deepmaps: Reading Tim Robinson's Maps of Aran," *Terrain Org: A Journal of the Built and Natural Environments*, Spring/Summer 2012. It has become an important metaphor for the field of spatial anthropology, which seeks to understand human activity in terms of the places in which it occurs. As one delves deeper into a place successive layers of meaning emerge, providing ever richer pictures. See Les Roberts, "Deep Mapping and Spatial Anthropology," *Humanities* 5, no. 1 (January 14, 2016): 5.

21. https://digital.library.nuigalway.ie/islandora/object/nuigalway%3A10957.

22. Tim Robinson, *Experiments on Reality* (London: Penguin Ireland, 2019), 118.

23. Tim Robinson, *Connemara: Listening to the Wind* (Dublin, Ireland: Penguin, 2006), 3.

8

What Is a Map?

> The periodic table, engineering sketches, business graphs, timelines, musical scores, sentence diagrams, Venn diagrams, circuit diagrams, and computer "desktops" can all be considered maps of a sort. Each of these devices takes one kind of information and "maps" it onto two-dimensional images. . . . A geographical map may be the most intuitive kind of graphic communication device, because it uses special relationships on the page to represent similar arrangements in real space.
>
> —Gary Benenson and James L. Neujahr, *Mapping*[1]

We all know what maps are. They show up on our phones when we ask for the location of some place. They tell us the weather, or increasingly, it seems, the spread of fires and floods. We see them in news articles, showing us the location of the latest global trouble spot. Some maps appear in board games. Others reside in the rarely used atlas up in the attic.

Some of us fantasize about finding a colorful map on thick, yellowed parchment, with figures representing palm trees and sailing ships, some vague directions, and an "X" marking the spot for buried pirate treasure. Maps like that are themselves hard to find and, invariably, hard to follow. What we eventually learn is that there is no buried pirate treasure, but that old maps are themselves a treasure.

A bestiary of maps would be large and diverse. Even traditional cartographic maps are not as simple as we might think at first, nor do they fit into a single well-defined category. Maps are more than just these odd objects; they are an essential aspect of thinking. They are forms of art, tools for science, tools for analysis, and an ever-evolving form of technology.

For example, a teacher and children created a map to record the temperature in a large drafty classroom (figure 8.1). The contours represent different temperatures. As Benenson and Neujahr say, the map raises interesting questions: Where is the heat coming from? Where is it leaking out? Are any of the windows open or broken, or are they just old windows that don't keep in heat very well? These could guide a path toward a solution, or possibly call for revision of the map or creation of alternative maps.

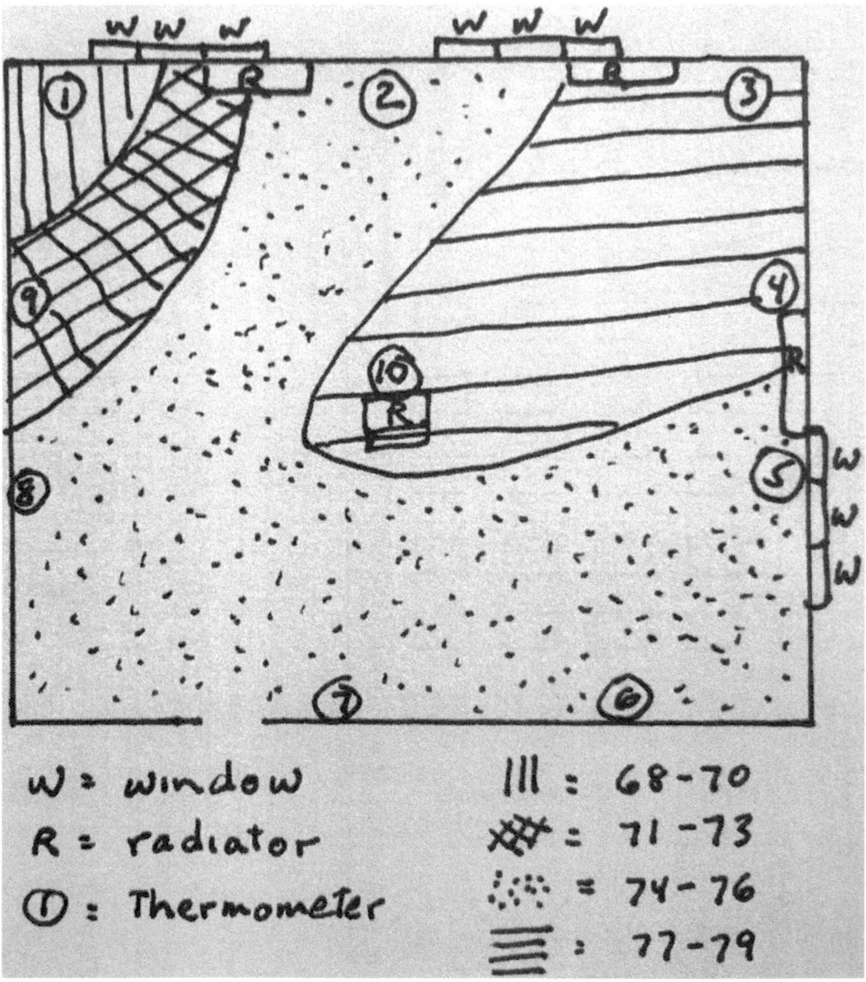

Figure 8.1. Contour map of temperature distribution in a room. ***Gary Benenson and James L. Neujahr,*** **Mapping** ***(Portsmouth, NH: Heinemann Educational Books, 2002), 25***

Frank Jacobs opens *Strange Maps*[2] with the observation that cartography is bounded only by our imagination, the element that strikes our curiosity:

> It might be the rarity of their subject matter: cartographic misconceptions, ephemeral states forgotten by history, bizarre idiosyncrasies of certain boundaries. The striking element might also be the treatment: the wild cartographic exaggerations of parody and propaganda, the skewed perspective of artists, maps forming the necessary background for literary invention. Or it might be in the method: statistics masquerading as maps, revealing some information about economy, culture, linguistics and so on. It might be anything.

This chapter explores the very definition of maps. In the process it considers how maps are tools for thinking and learning in every imaginable area.

MAPS AS THINKING TOOLS

Most of us rarely think about maps, and many people deem their usefulness to be limited. We are not born knowing how to read them or to create them, as is evident when we ask a stranger for directions. We may avoid maps, especially because they often use such obscure symbols as the one for occluded fronts on a weather map.

However, maps have long been a tool for thinking in every discipline, much as writing or mathematics can be. They appear in every culture in ancient times.

Figure 8.2 shows what may be the first city map from circa 1500 BCE. It is on a clay tablet, about 18 by 21 centimeters. The map shows the city of Nippur, the religious center of the Sumerians in Babylonia. It depicts the temple of Enlil, a city park, storehouses, a park, the river Euphrates, canals, and the city wall including its gates. It may have been created to guide repairs of the city's defenses. It is the earliest known town plan drawn to scale and showing orientation.

Today, we have maps of the human genome, evolutionary and ecosystem relationships, the economy, musical compositions, historical forces and changes, human migration, and much more. Early examples, for instance, the Nippur tablet, show how maps have long entered into both complex calculations and everyday life. They remind us that the role of maps in thinking and learning goes well beyond simply learning how to use a map or how to memorize place names. Understanding that role can help us to support education that nurtures experimentation, creativity, critical thinking, and humanistic understanding.

Figure 8.2. Clay tablet map from Nippur, Sumeria (now in Iraq), ca. 1500 BCE. ***Public domain***

CLASSICAL EDUCATION AND MAPS

Many universities have closed their geography departments, or merged them with history, urban planning, geology, or other disciplines. Schools, which once taught geography as a major subject, now slip it into social studies, and often there in the briefest fashion. This leads many people to think of it as an antiquated subject, no longer needed in the era of smartphones.

The perception that geography is antiquated has some validity. In a medieval university the way, or road, to knowledge was through what we would call geography. It had two sets of lessons to learn. The first was the *trivium* (from Latin, "the place where three roads meet"). This was the introductory curriculum—the study of grammar, rhetoric, and logic. These fields constituted what we might call the language arts.

Following the trivium was the *quadrivium* ("four roads"). These were the mathematical arts of arithmetic (number), geometry (number in space), music (number in time), and astronomy (number in space and time). Geometry meant not simply examining abstract postulates about triangles and parallel lines, but also measuring and comprehending the actual Earth. Astronomy was studying the heavens through numbers. Arithmetic meant numbers, seen as the key to the universe, and music was number in time. This did not imply formal study of the kind seen today in those disciplines, but their use as tools for understanding the world. Especially through geometry and astronomy, along with an expanding array of explorations and travel writings, the quadrivium stimulated the emergence of geography.

The trivium implied learning language, learning other things through language, and learning about language. In a similar way, the quadrivium meant learning geography, learning other things through geography, and learning about geography.

The modern educator might gasp, "Where are computer science, biology, chemistry, physics, civics, literature, history, and so much more?" The medieval educator would counter that before learning many other things, including the practical arts, for example, medicine, it is important to have a grounding in the world we live in and our place in it.

With advances in pedagogy, we no longer follow the three or four roads in the same way, but we may have lost something in the desire to cover, or rather touch lightly upon, so much in the modern curriculum. We especially seem to have lost the idea of geography as a core capacity that enables learning many other things.

Fashions in education change, but maps are still vital for human experience. Before delving into all of the reasons why they matter, examining the different types of maps, exploring how maps are crucial for thinking and learning, discovering why we still manage to get lost even using our thousand-dollar phones, and questioning why the discipline of geography is in crisis, it will be useful to ask, What is a map?[3]

PHYSICAL REALITY

The International Cartographic Association (ICA) defines a map as follows: "A map is a symbolized representation of geographical reality, representing selected features or characteristics, resulting from the creative effort of its author's execution of choices, and is designed for use when spatial relationships are of primary relevance." The ICA definition includes some important features of maps. One is that there is a spatial realm of some interest and that it is possible to produce a "symbolized representation" of the elements within it. Presumably the map is easier to use, more accessible, or more revealing than direct access to the space itself.

Also, any map is a "creative effort," sometimes requiring major aesthetic decisions and inventions. Even an aerial or satellite image is a constructed symbol that manifests decisions about focal length, angle of view, coloration, and other features. The definition works well for one major category of maps, *reference maps*:

- *General reference (or planimetric) maps* show important physical (both natural and man-made) features in an area. They can display geographic boundaries, physical features, or cultural features. Examples include political maps, road maps, topographic maps, and geologic maps.

When one considers actual information ecologies of use, it is clear that we need additional categories. Apparently small differences portend large differences in the details of maps and how they are used. Australia's Intergovernmental Committee on Surveying and Mapping (IGSM) singles out three additional kinds of maps:

- *Topographic maps* are like general reference maps, but they show elevation in detail. A standard technique is the use of contour lines, which join points of equal elevation above sea levels. They can reveal characteristics of landforms and terrain, lakes and rivers, forest cover, and grasslands, as well as human-made features, for example, administrative areas, population distribution, and transportation—air landing fields, roads, canals, and railways. Topo maps are often produced by government agencies, especially for defense purposes.
- *Navigation charts* combine aspects of the aforementioned maps. They are used as navigation aids for boats and aircraft.
- *Cadastral maps and plans*: A *cadastre* is a register of property titles, managed by a government agency. The registry comprises *cadastral plans*, each showing a description of a parcel of land, its location, and who owns it. It may also record what the land can be used for (e.g., residential or commercial, conservation), the location and shape of buildings, and the value. *Cadastral maps* are made up by joining together a set of cadastral plans (horizontal alignment).[4]

A major tool for organizing information on a website is the spatial organization implicit in general maps. In this case, the entries, which can be called *place descriptions*, are organized by their *geotag*, or *where* they occurred. For example, a Web page about the Frisian Islands might be more about the place than about a visit at a particular time. Just as pages can be grouped into a tree structure or posts into a chronology, place descriptions can be grouped into a *spatial map* using their geotags.

The definition also expresses commonsense notions of maps, which we examine in what follows. One such notion is that maps represent geographic reality. That notion has two problems. First, all geographic maps construct a portrayed reality, which may vary greatly from other conceptions of that reality. The ICA opens that door with its emphasis on "representing selected features or characteristics." By choosing particular features, say mountains, the mapmaker seriously compromises full and accurate representation. For some uses this may be a useful maneuver; for others it can be disastrous.

Although that definition might describe many maps, it is not very helpful, because "geographic reality" covers a lot of ground. But that is also the power of mapping: it is a thinking strategy that applies to almost everything we do. Such terms as *geographic reality* or *spatial relationship* are difficult to define, that is, unless one uses a map.

In some cases, the mapmaker has limited information. Think of pre-Columbian mapmakers who left out the Americas in their world maps. In others, the mapmaker deliberately presents the world in a way to further some rhetorical purpose, which may deviate strongly from "reality," or, we might say, constructs reality in a particular way. Think here of the political maps that arose after the First World War in the Balkans and the Middle East. The Sykes–Picot Agreement and various other treaties led to maps that attempted to define a reality that the world struggles with to this day.

Second, the conceptual distinction between maps used to show "geographical reality" and those used to represent other realities is not sharp. The ideas of "representing selected features or characteristics" and "creative effort of its author's execution of choices" allows for representing features that may not be visually apparent.

INFORMATION ON A TOPIC

A cartographer might also want to include indications of capital cities, population, or other such information that is geographic in the broad sense, not simply a portrayal of physical reality. Geographic reality could include species distributions, musical and linguistic practices, family income, or essentially any other information. In the end, spatial relationships are not "of primary relevance" for many of these concepts, yet they could easily be part of useful maps.

Thematic Maps

The idea of displaying a topic and its geographic variation leads to another major category of maps:

- *Thematic maps* show the variation of a theme or topic across an area. These include maps of weather, income, resources, population density, and geology.

Thematic maps are an example of combining a portrayal of a physical space, for instance, a country, with a symbolic portrayal of some theme, for example, weather features. These thematic maps are essentially *overlays*, or vertical alignment of two or more maps, one focused on spatial relations and a second focused on a theme or topic.

Some thematic maps show continuous distributions, for example, population (of people or other species), income, disease prevalence, or rainfall. Some simply display a theme in terms of characteristic images, for instance, state birds arrayed on a U.S. map.

Life magazine created an important example of a thematic map for its first issue of 1937. This was near the end of Franklin Roosevelt's first term as president, when many people wanted to celebrate the public works projects, and some complained about the cost. In this context, a map of the United States is spread over two full pages. The headline reads, "What President Roosevelt Did to the Map of the U.S. in Four Years with $6,500,000,000."[5]

Across the map are images of projects from FDR's first term. One is a dam, which created the Pocatello Reservoir on the Snake River in Idaho. The image for the dam is displayed at nearly the size of the state, so its location is only approximate. Similarly, on the map are a high-rise housing complex in Indianapolis, the Philadelphia Naval Hospital, and Manhattan midtown tunnels.

The images are broad strokes indicating the idea of each project, rather than its precise appearance or location. One for New York shows newly planted trees near a lake with the caption "forest conservation." Some images have no single location, for example, "120 U.S. airports," "Lake Okeechobee flood control," or "2 U.S. aircraft carriers, 50 U.S. warships."

A typical reader would certainly consider this montage of images and map to be a map, one that displays where projects occurred, even if closer examination shows that not everything is place related. The general impression is of a diverse array of projects spread across the country and, of course, the high cost for its time of $6.5 billion.

A thematic map without its accompanying portrayal of a physical space could still be useful. In fact, it moves us in the direction of general visual displays of information. For example, data on income/poverty levels could be represented in a histogram, essentially a map of that information.

In that case, is spatial relationship still a criterion? It is if it has an explicit—or even assumed—spatial dimension, for example, "in North America." But the histogram per se, a map in every other way, and embodying implicit geographic relations, might not look anything like a conventional geographic map.

These considerations suggest that, for our purposes, we need a different definition. An alternate way to define a map is to note that any map is itself a space. It uses spatially distributed elements (up/down, left/right, forward/back) to represent elements in some system and to show the relations among them. That system could itself be inherently spatial or may simply benefit from a spatially distributed depiction.

A Map as a Display of Information Using a Spatial Metaphor

This flips the story. Instead of a map being a representation using symbols on paper to depict a physical space, the map is itself a space, in which its spatial properties can be used to represent relations in the world. Making this flip allows us to consider representing elements and relations that may in themselves not be spatial.

The website 100 Amazing World Maps by Lissa Poirot shows examples of thematic world maps (many of the United States) on a variety of topics: a "Cartoon Map of Europe, 1914" depicts how Germans perceived geopolitical threats, especially Russia. Each country is represented by a human cartoon. "Map of Rivers Flowing into the Mississippi River" shows the extraordinary number of rivers in the Mississippi River basin. "Map of the World's Lactose Intolerance" shows worldwide distribution of lactose intolerance with numerical percentages. "Map of Water Consumption around the World" uses a planimetric world map along with pictograms, bar charts, and other symbols to report on water use.

The purpose of maps like these is not to depict physical reality precisely but to use that as the basis for displaying some other information, which can be linked with physical location. The spatial display, including relative sizes of regions, becomes a language for expressing that information.

A broader definition that captures these purposes could be the following: "A map is an object whose spatially distributed elements show the relations among elements of some system." One advantage of this new definition is that it does not require the elements in the system to be themselves in a spatial relationship, only that the relations among them can be illustrated using spatially distributed elements. Thus, the world represented in the map does not itself necessarily exhibit spatial relations, only its map.

Although that definition might describe any map, it is not very helpful, because "some system" covers a lot of ground. But that is also the power of mapping: it is a thinking strategy that applies to almost everything we do.

Networks of Concepts

How do we find an entry on a website? Usually, we do it either from a general search or by a link from another entry. Sometimes we follow a link within one page to find another. All of these are enabled by the fact that the entries are connected in some pattern.

One such pattern depends on *what* the entry is about. When we focus on the *what*, we call the entry a *page*. The main page on a topic (the *parent*) then links to subtopics (the *children*). An example of this is the Wikipedia page on art, which has lengthy sections within it, for example, "Creative art and fine art," and also links to many other pages, for instance, "Imagination," or specific artists. From the perspective of the main entry on art, all of these are its children pages.

Another example is an organizational chart that turns the hierarchy of an organization into a spatial display, typically one in which "up" means more important, for example, a board of trustees. The spatial display on any map gives us a powerful way of abstracting from the complexity of any domain. It relies on our general capacity for spatial reasoning and the seemingly endless ability to convert any set of objects and relations into a spatial display.

There can be many generations of pages, resulting in a complex family tree of pages. Of course, with hyperlinking, it is not a strictly hierarchical tree structure but a general graph. However, the fundamental idea from the perspective of any one page is not that different from the kinds of outlines we were taught to do in school.

A webpage *site map* is, then, a map of concepts, not a map of physical reality, for example, a road map. It reminds us that concepts, for instance, physical features of a landscape, can be mapped as well. In fact, the line between concept maps and conventional physical maps is often blurred, especially with thematic maps or multiple maps showing change throughout time.

Spatial, Temporal, or Conceptual?

A subway map provides an interesting test case for the idea of concept maps. In one sense maps of subway systems, rail lines, ferry routes, and other such systems appear to be a standard case of representing spatial relations on a planar surface, like any general reference map. But the spatial distances are often distorted to simplify or clarify the display. Three nearby stops might be separated widely for readability, maintaining only their sequential relationship. Similarly, a faraway station might be shown as just the next stop without trying to be true to the actual distance.

The result is that the subway map represents the conceptual relation of sequence, which may bear little relation to the miles traveled. If anything, the map represents travel time over distance. Going through eight stops may take significant time with stopping, unloading, loading, and even delaying to match a schedule, whereas one long uninterrupted stretch might be traversed rather quickly.

This means that the spatial world of the subway map is dramatically different from what a precise reference map might show. It is a concept map evolved from a spatial display, which might equally be displayed through a network of timelines.

Cladograms

The great diversity of living things has always called for some kind of grouping or classification. We may not be able to distinguish one mushroom from another, enough to be confident about eating it, but we feel confident that the mushroom belongs in a different group from the oak tree, and certainly from a butterfly or a horse. How can we make sense of the amazing variation among species, including both the variety and the similarities?

Many methods have been developed to do just that, and many corresponding maps. For example, a *cladogram*[6] show relations among organisms (though not how ancestors are related to descendants). Lines branch off from a common trunk, ending at a *clade*, a group of organisms with a last common ancestor. The hypothetical ancestor may provide clues about the order of evolution of various features, adaptation, and other evolutionary narratives about ancestors. Although traditionally such cladograms were generated largely on the basis of morphology, molecular data are now commonly used.

Figure 8.3 shows a cladogram for Endopterygota, the most diverse insect superorder, with more than one million living species. The highlighted order, Trichoptera (caddisflies), alone has 14,500 described species, possibly many more yet to be classified. A complete cladogram for even that one order would be overwhelming. However, the abbreviated cladogram allows us to see (or postulate) relationships and understand connections better.

The American Museum of Natural History, located on Manhattan's Upper West Side, is one of the largest museums in the world. Its complex comprises twenty-seven interconnected buildings housing forty-five permanent exhibition halls, plus a planetarium and a library. Many maps help in navigating this large space.

On some floors, a map of exhibits is in the form of a cladogram. As visitors explore the branches, they see examples of creatures in each clade. Their own body becomes an investigation tool, so that greater distances between exhibits correspond roughly to greater evolutionary distance between organisms.

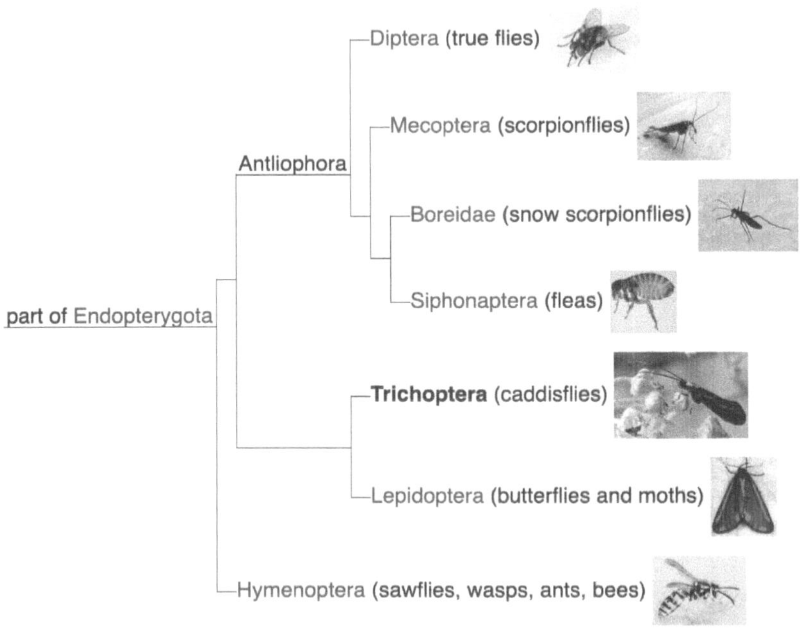

Figuire 8.3. Cladogram for the Endopterygota superorder of insects, based on DNA and protein analysis. *Wikipedia, https://en.wikipedia.org/wiki/Caddisfly*

Creating Concept Maps

Students probably learn more from creating their own concept maps than from reading preconstructed ones. For example, a group of American college students collaboratively developed one on a whiteboard not only to express, but also to develop their understanding of "the essential developments and events of European history since 1789." Their map emphasizes World War I, communism, and the European Union. The map has multiple centers and, thus, invites different narratives based on the concepts identified.

Sometimes a concept map can be combined with a traditional map of physical features. A commentary in *Nature* presents mounting evidence that climate tipping point events due to CO_2 emissions and consequent global heating could be more likely and have more severe impact than previously believed, and that they are interconnected across different biophysical systems. This potentially commits the world to long-term irreversible changes.[7] The authors make their case in text and with a cause-effect concept map (figure 8.4). That map is overlaid on a global 2-D projection showing the location of the tipping point events. Thus, the map is both concept map and traditional global map.

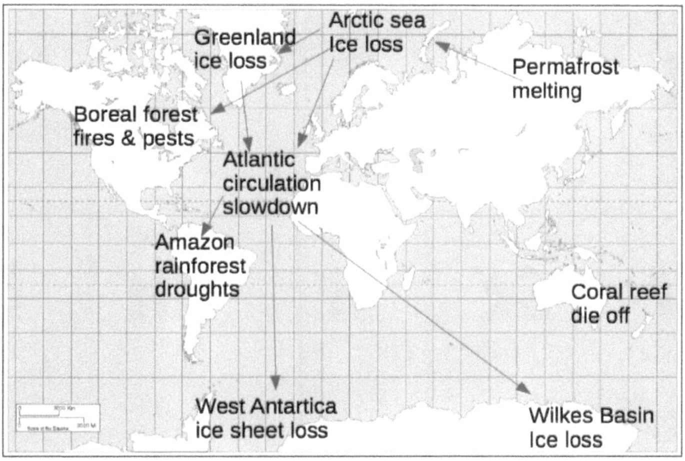

Figure 8.4. Potential climate tipping points, with causal links. ***Author, adapted from Lenton et al., "Climate Tipping Points—Too Risky to Bet Against," 2019.***

A Zoo of Visualization Methods

Once we realize that maps can be used to represent in a spatial way concepts and relations of any kind, the realm of visualization methods opens up. Ralph Lengler and Martin Eppler present a typology for visual displays of information and use it to construct a periodic table of visualization methods, with more than one hundred types.[8] A small portion of that table is shown in figure 8.5. The first two columns show methods for *data visualization*; the next three are for *information visualization*. Other symbols indicate convergent versus divergent thinking (the examples all happen to be convergent), overview or detail, and other features.

Their typology includes *data visualization*, for example, with histograms, pie charts, and scatter plots; *information visualization*, for instance, semantic networks and tree diagrams; *concept visualization*, such as concept maps or Gantt charts; *metaphor visualization*, for example, a story board; *strategy visualization*, for instance, flowcharts; and *compound visualization*, combining several other formats.

Although their typology seems at first removed from IGSM's world of maps, it is easy to see from actual examples how closely these coordinate. For example, a *compound visualization* combining several visual display formats is akin to a thematic map, which combines geographic information with some theme.

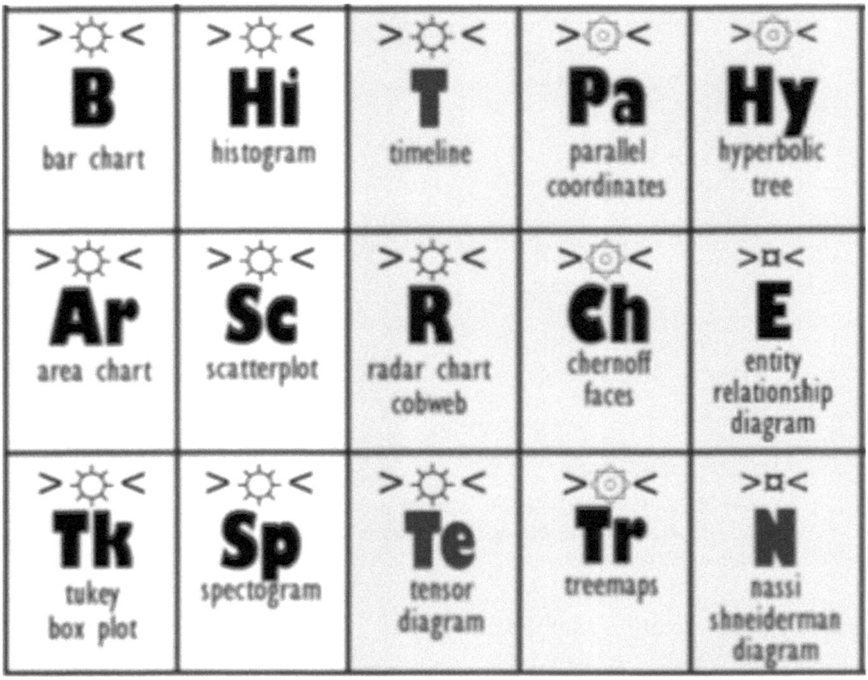

Figure 8.5. From the periodic table of visualization methods. ***Author, adapted from Ralph Lengler & Martin Eppler,*** **Proceedings of the IASTED International Conference on Graphics and Visualization in Engineering, 2007**

PHYSICAL MANIFESTATIONS

When you imagine holding a map, that might be a familiar paper map, an atlas, a globe, or some digital variety on a computer screen or a mobile device. But although these objects seem ubiquitous, they are but a small portion of the territory of maps. The physical manifestations of maps could easily fill another book; they appear in every conceivable medium.

Long before modern sea navigation charts, sailors in the Marshall Islands learned to navigate through stick charts, known as *rebbelibs* (figure 8.6). They are made of bamboo sticks and cowrie shells arranged across a parallelogram frame. The sticks represent ocean currents, wind patterns, and wave swells. The cowrie shells represent island groups, not individual islands. They serve as navigational charts for the Marshall Islands region.[9]

For example, Columbus Globes makes floor-stand globes, desktop globes, mini-globes, student globes, lighted globes, and more. They once made a "Geo Disk." It was a twelve-inch diameter, stiff paper device, with a rotat-

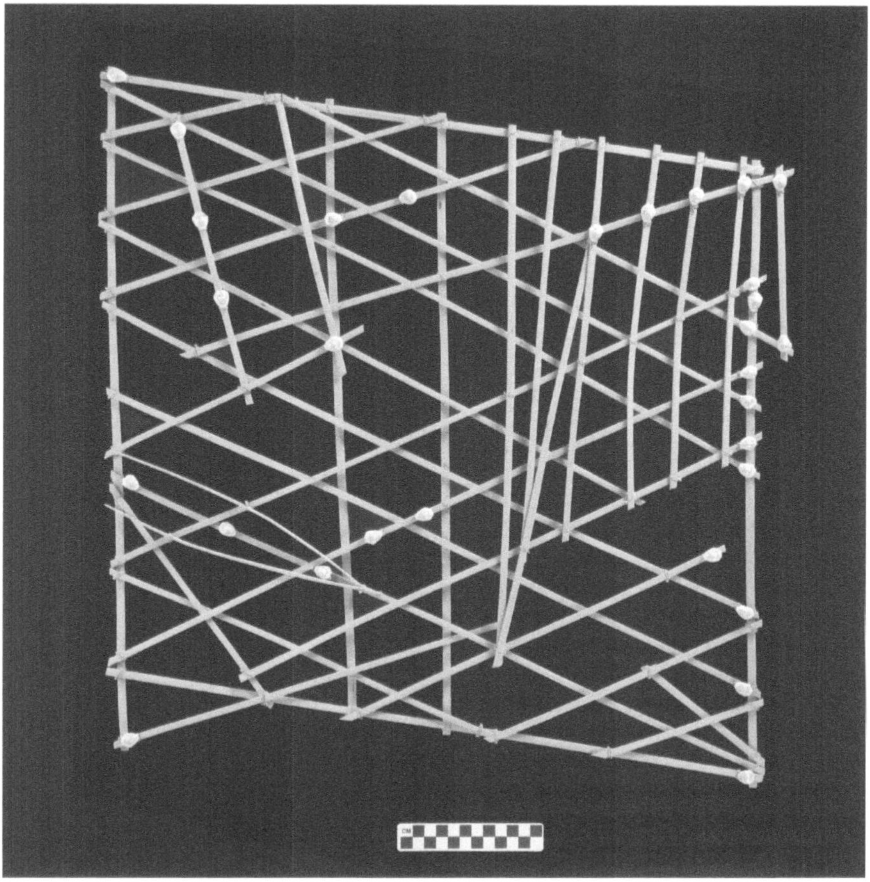

Figure 8.6. A stick chart (*rebbelib*), made of bamboo sticks and cowrie shells arranged on a parallelogram frame. ©*Denver Museum of Nature and Science*

ing "info tool" to look up country names, capitals, population, area, and languages. It had a world map and all the country flags in a compact form. Similar "dist-o-maps" were once common. The user could rotate them to find the distance from one town to another.

A map for the Pyrénées/Pironeos in the form of a book with continuous cardboard pages folds out to fifty-one inches. One side shows the French side of the Pyrenees and the other the Spanish. The top of the map is cut to show the peaks, roughly to scale. It shows towns, rivers, and caves, but interestingly, no roads or national boundaries, such as for Andorra. It has illustrations of local wildlife, culture, and historic sites. Is this a book? A poster? It is surely a map of some kind, but one that is rarely seen elsewhere because its form so perfectly matches the three-hundred-mile roughly straight line of the crest of the Pyrenees.

Maps of portage (or carry) trails in canoe country indicate where one can carry a canoe from one lake to another. The maps exist not only on paper and digital form, but also in the form of signs at the start of each trail and blazes along the way. Those trails are sometimes maintained by environmental or recreational organizations, but the modern trails build upon centuries of use by other canoeists, including adventurers, soldiers, traders, trappers, and settlers. Many such trails were created/mapped even earlier by French voyageurs. Their trails, in turn, were built upon those of indigenous people, who wisely relied upon even earlier paths that moose, deer, and other wildlife created.

The stations of the cross give Christians a map to follow, one that tells a story. They usually comprise a series of fourteen numbered images arranged along a path. The faithful travel on the path, stopping at each station to say selected prayers. In some cases, the stations have small plaques with bas-reliefs or paintings and are placed around a church nave. Others are outdoors, often with widely separated stations, but still constitute enough of a map that a newcomer can find his way through all fourteen stations. Sometimes they use statuary. When the pope leads the stations of the cross around the Colosseum on Good Friday, the faithful might experience and say the stations without there being any image.

Other maps appear in virtual or augmented reality; on road signs; in videos; on curtains, rugs, and wallpaper; on postage stamps; on subway displays; even on tattoos. A map on the wall or in a brochure can be read and followed (e.g., through a museum or a zoo); the same map could appear on the pavement itself. It could be a map with raised images or braille script for visually impaired people. Even the smells of the zoo or other site could provide a map. Imagine the smells of a bazaar—fresh fruit, spices, fish, and barbecue.

It is no accident that many people who write about either maps or walking conclude by virtually equating the two. One can see this in the way that Rebecca Solnit comments on how the bodily act of walking defines our world: "The path is an extension of walking, the places set aside for walking are monuments to that pursuit, and walking is a mode of making the world as well as being in it."[10]

Walking thus becomes an artform: "the way rural walking measures the body and the earth against each other, the way urban walking elicits unpredictable social encounters . . . the way walking reshapes the world by mapping it, treading paths into it, encountering it; the way each act reflects and reinvents the culture in which it takes place."[11] Our footprints, or the sidewalks we later build, tell us and others where we have walked and may go in the future. Note that a walk provides a multisensory experience lacking in most maps.

The global Geographic Positioning System (GPS) uses a network of satellites and receiving devices to determine a location on the Earth's surface. Some GPS devices are accurate to within one centimeter. GPS is usually described as a tool to use with a map. It facilitates correlating a current location with a place in the region that the map defines. But one could think of the GPS itself as a map. Rather than a visual display, it offers access to a coordinate system: any place can be identified by its coordinates using a GPS receiver; and knowing the coordinates, one can find any place. This is something that many would expect the ideal map to do.

Building on GPS, the advent of spatial computing has brought maps into the center of the digital age. Spatial computing means many things, but a good starting point is that it is "human interaction with a machine in which the machine retains and manipulates referents to real objects and spaces."[12] Sensors, including cameras that record in multiple wavelengths and sensitive microphones, connect the machine to the physical world once reserved for living intelligences. Many see spatial computing as a new frontier for technology development, essential for serving human purposes in work and play.

Medical researchers use it to track diseases in an extension of John Snow's work. For example, Chang et al. used cell phone data to map the hourly movements of millions of people from neighborhoods to such points of interest as restaurants and religious establishments. Incorporating daily case counts and epidemic models, they identified COVID-19 "super-spreader" points of interest. Knowledge derived this way can be invaluable for public health planning. A practical finding is that restricting maximum occupancy at points of interest is more effective than uniformly reducing mobility.[13]

The very power of spatial computing has attendant risks. In situations of natural disasters, for instance, hurricanes, fires, or a pandemic, people rely on popular disaster apps to identify whether they are in harm's way and to monitor the overall extent and severity of the disaster. The app essentially encodes a map that combines general data about the disaster with individual data, for example, the immediate location, home address, or personal condition of a user. The user becomes part of the map.

In these situations, people typically relax their usual privacy expectations. Recent studies have shown large gaps between privacy practices and information flows, and between governance and practice.[14] Data collected for the emergency is often maintained long after it is needed and not sufficiently protected. This applies to disaster apps, proximity trackers for disease exposure, and other otherwise useful devices.

Scientists across a variety of disciplines use spatial computing to map the surface of Mars or the bottom of the ocean, to track migratory species, and monitor climate changes. The internet of things, in which devices from doorbells to speakers to home appliances "know" where they are, promises new ways of living dependent on machines in convenient but unpredictable ways. Modern cars alert drivers to dangers, indicate the route, and suggest places to visit; driverless cars go a step further, using spatial computing to take riders to their destinations.[15]

It is worth noting that not all maps have physical manifestations. Some exist in our dreams, our imaginations, and our plans for the future. For example, a Halloween trick-or-treater could envision a map of houses in the neighborhood likely to have good candy and a circuit to visit them all. A shopkeeper might imagine a cabinet with drawers to hold items for sale, some of which currently are in inventory and others only contemplated. That cabinet would be a memory aid as well as a map of objects grouped into useful categories.

MAPS OF TIME

Pages on a website are typically organized by concept hierarchies. An alternative is to focus on the *when* of an entry. In that case, the website becomes a *blog* (or *vlog* for video-based blogs) and the entries, now called *posts*, are organized by their time of creation into a *chronology*, usually with the most recent first. Websites can thus map information by time.

But time can also be indicated by timelines and other special types of maps. One commonly used approach is to create a series of maps: for example, maps of a city as it changes throughout the decades. Consider also explicit timelines, wikis, time-sensitive bar charts and line graphs, and automatic construction of chronologically organized Web pages. In *Cartographies of Time*, Daniel Rosenberg and Anthony Grafton write,

> The timeline has never been more important or more ubiquitous than it is today. . . . Along with the list and the link, the timeline is one of the central organizing structures of the contemporary user interface . . . the sheer volume of information now readily available in electronic form puts indexing systems of all sorts at a premium, and the availability of dynamic, animated forms of presentations places special emphasis on time is an axis of organization.[16]

The Geologic Time Spiral (figure 8.7) shows recent world time periods as a sort of curving timeline, then suggests the vastly greater pre-Cambrian times as a spiral receding into the distant past.[17] If everything were shown to scale, the illustrated portion would be only 10 percent of the spiral, and the detail would be lost.

Figure 8.7. The Geologic Time Spiral: A Path to the Past. ***Public domain, U.S. Geological Survey***

FUNDAMENTAL LAWS OF MAPS

We can define maps better by articulating some of the *fundamental laws of maps.* These may be grouped in terms of purpose.

Purpose

- The creator or constructor of the map has a rhetorical purpose, just as the writer of a text seeks to persuade, inform, or entertain the reader in some way. Every map is designed with some use in mind, although it may satisfy other purposes later on.
- It is often useful to make the rhetorical purpose explicit. For example, the mapmaker could guide the user to focus on the details in a complex map

that express its main point. In other cases, most notably in marketing and political maneuvering, the mapmaker may hide its underlying purpose.

- The interpreter of a map may impose a purpose different from that of the maker. For example, a tourist map might be designed to highlight businesses in a community, particularly those that paid to be included. A pollution investigator could use the same map to ferret out possible sources of water pollution.
- The issues of purpose are inseparable from those of power: Who gets to make the map? What aspects do they emphasize? How do they draw the lines? What is their point of view? Maps connect us to realms of property lines, voting rights, taxation districts, and enterprise zones, which would not exist without the map.[18] For children in school, for indigenous people, for those marginalized in some way, power defines which maps get created or used, and how they are interpreted.

Diversity

- The diversity of maps is as great as the diverse purposes or questions that anyone may have. Any area of cognitive activity offers a potential for maps to facilitate problem solving. It does this by abstracting from reality, whether geographic or other, to allow some phenomenon to be examined, modeled, and manipulated.

Alignment

- There must be a way to *align* or *register* the map with the elements and relations it depicts. This could be done, for example, by showing boundary markers (or boundary stones) that have an actual physical manifestation. Maps of urban areas can enable registration by identifying street corners or major buildings.
- Maps are more useful when they align with other maps as well. Maps can align vertically with overlays showing multiple planes of information. They also align horizontally by abutting other maps.[19]

Construction

- The elements depicted on a map are themselves constructed. For example, some maps may show trees and others not. But what is a tree? Does the concept include telephone poles made from trees? Dead trees? Saplings? Bushes? The map constructor must define "tree," then decide which elements in the space to portray, and in what way.

- Because the map constructor decides which elements to portray and how, a map can be useful for any elements in any space. The familiar realm of maps of objects on the land needs to be expanded to include the sea, the solar system and galaxies, cell structures and DNA, histograms, pie charts, sentence diagrams, organization charts, words, and ideas. Maps can make small things larger and large things smaller. We do not need a metaphor to talk about a flowchart or a mind map as a map; maps are already symbolic portrayals of conceptual objects, not one-to-one depictions of physical reality.
- Maps require specifications, for example, *orientation* (is north at the top?), for their construction, which help them achieve effective use.[20]
- Specifications also include *data definitions* and *pictorial conventions*.[21]
- The usefulness of a map is enhanced when the specifications, as well as purpose, are made explicit.

Maps and Thinking

- Not everything is a map, but maps are essential in our relation to everything we think about.
- A map is a *model*, representing selected aspects of the world. It uses *symbols* to convey information. In this way, maps are similar to both written text and mathematical expressions. Each of these systems allows us to simplify a complex world to understand it or act upon it.
- The three systems (text, numbers, maps) may operate in parallel, but they also intermingle. Maps usually include text and numbers, just as the other systems often rely on or incorporate maps.

These laws of maps lead to an expanded definition, closely akin to saying a "visual display of information":

> A map is an object whose spatially distributed elements show the relations among elements of some system. (It is designed, or appropriated, for use. Ideally a map includes specifications of its display conventions, and the potential to align both with system being portrayed and with other maps.)

THE MAP THAT IS NOT

The last formulation of a definition for maps has an odd consequence. What may be the world's most commonly used map (or set of similar maps) functions in the most limited way as a map.

When someone says that she needs to check a map on her phone (or tablet, laptop, etc.), she usually means that she wants to use a Google map, an Apple map, an OpenStreetMap, or other such resource embedded in a digital app. She typically does so to find some place, or to determine the fastest route. These apps are very effective for that purpose, including the option to hear audio directions, for instance, "Turn left in five hundred feet." This is probably the most common use of maps in the world.

Some other people, even if they occasionally use a map app, argue for the value of road maps or atlases. A bulky paper-based tool somehow seems more useful at giving the big picture. That picture, in turn, can teach something about the region. It can suggest alternative routes, for example, one that goes near a state park with a waterfall, or another that goes through a medium-sized town that might have an old-style family restaurant for lunch. An argument about new versus old maps then ensues.

But in a sense, the map app is not a map at all. At least, as most often used, it does not assume many of the characteristics of maps. It is not designed to "show the relations among elements of some system" but to provide an easily understood solution to an immediate problem (e.g., which street do I turn on). Often, "specifications of its display conventions" are limited, for example, scale and orientation, or one has to put in some effort to bring them up. A map app has little "potential to align both with system being portrayed and with other maps."

FINDING OURSELVES

When using the map on a smartphone, it is easy to find yourself out of the picture. You may see roads, highway interchanges, towns, tourist attractions, and such but not know where you are relative to them. This is a *registration* failure: the display leaves you disconnected from the real world. This may signal a good time to turn to the re-center button, which places at least a representation of you back in the center of the action.

You may still be lost. If the scale is too large or too small, the map may not provide enough useful information to orient yourself. You need to register or match up the symbols in the map with the objects you know or can see from your vantage point. Again, technology has a fix. You can use swiping gestures or the "+" and "–" to adjust the scale. Similarly, technology fixes will find north again, or reveal your desired destination.

The lack of these features is not a defect. The user actually does not want a lot of information that might distract from the task at hand. He may literally have only a second or two to make a decision about a lane change. In fact, he may not want "an object [with] spatially distributed elements" but simply a clear command. That is why the audio component is so helpful.

For these reasons, any debate about paper versus phone is pointless. The distinction is not one platform for a map versus another, but use of a map versus use of a route-guiding tool, which looks a little like a map, and is called a map, but it is not much of a map in typical practice.

Of course, digital tools are also used for general map purposes, for instance, gaining an overview of a region, opening up to serendipitous discoveries, gaining aesthetic enjoyment, recording a journey, and such. Some of these functions could be added to a phone map app, but too many of them would defeat its primary purpose of route guiding, which is the function you seek when you do not really want a map.

TOOL MAPS

Maps represent all sorts of information that we might wish to learn. Often, the visual display of a map makes it easier to comprehend that information, especially when it is novel or complex. It might require a lengthy discourse to convey the same information in linear text, if that could be done at all.

The map is, thus, a more effective way to convey the information, and it also provides a means to remember it, apply it, or convey it to others. For example, a budding chemist can use the visual display of Dmitri Mendeleev's periodic table to recall atomic numbers and weights, valences, the "inert" gases, the metals, the trans-uranic elements, and much more. Most of this information could be expressed in linear text or using numerical tables, but even without the table, the successful chemist would undoubtedly need to devise some visual scheme to remember it all.

Maps of this sort are reference maps. They provide detailed representations of information across many domains, often developed by the top experts in a field. As such they are indispensable tools for learning. But like the best books, these reference maps are most often created by others and afford only one sort of opportunity for the learner to make new meaning. Can there be a way for the novice in a field to learn as much as the makers of maps have learned through their construction process, perhaps as much as Mendeleev did?

Maps and Territories

Dewey's discussion of maps and territories is helpful here. He sees a (reference) map as a logical representation, which finds its parallel in psychological processes. An explorer may make a finished map after thoroughly exploring a country. Such a map would be based on the notes she made while blazing a trail in an unfamiliar area. The map and territory exploration are mutually dependent, but not equivalent:

> Without the more or less accidental and devious paths traced by the explorer there would be no facts which could be utilized in the making of the complete and related chart. But no one would get the benefit of the explorer's trip if it was not compared and checked up with similar wanderings undertaken by others; unless the new geographic facts learned, the streams crossed, the mountains climbed, etc., were viewed, not as mere incidents in the journey of the particular traveler, but (quite apart from the individual explorer's life) in relation to other similar facts already known. The map orders individual experiences, connecting them with one another irrespective of the local and temporal circumstances and accidents of their original discovery.[22]

The map can never take the place of an actual journey, just as the logically formulated material of a science cannot substitute for individual experiences. The map is an arranged and orderly view of actual or potential experiences. It can guide future experience, offering direction and control. As Dewey says, "it economizes effort, preventing useless wandering, and pointing out the paths which lead most quickly and most certainly to a desired result. Through the map every new traveler may get for his own journey the benefits of the results of others' explorations."[23]

"Map" as a Verb

In *Young Geographers*, Lucy Sprague Mitchell refers to reference maps as demonstration maps. They are valuable sources but meant to demonstrate some idea; they are to be read, not written. She takes Dewey's map versus territory distinction one step forward. Using existing maps must be complemented with the writing of maps, mapmaking: "More than demonstration maps are needed for the study of relations. A map is needed which could be actively used, not merely looked yet. Such a functioning map I have called a tool map."[24] The explorer learns both by using the maps made by other and by creating her own.

Mitchell's tool maps differ from reference maps in terms of both process and product. Their process is one of creating maps to enhance learning, not primarily reporting information for others. The focus is on making sense of relations in the world, not just display of already worked-out relations. Thus, "map" becomes a verb, an action to carry out, not a noun.

The real product of the process of working with tool maps is improved understanding, not necessarily the map itself. Benenson and Neujahr go further. They emphasize mapping as a process essential for dealing with a complex world. Maps are essentially models we create to work with complicated systems. The model, or tool map, is created to test ideas and relationships, only incidentally to be a definitive statement for the archives.

The idea of "map" as both noun and verb leads us to another definition, one that encompasses both reference and tool maps: "A map is a workspace in which the user employs spatially distributed elements to make sense of the relations among elements of some system." Mitchell expected that primary-school children would make tool maps out of maps, papier-mâché, cardboard, or drawings on paper. In general, they would not be altering the classroom's cherished globes or atlases. But with easy duplication of image files, inexpensive laser and 3-D printing, editable digital maps, and other tools, the line between professionally produced archival maps and learning maps that children make has blurred.

Modern cartographers often reveal their construction process and share the tools they used. Platforms like OpenStreetMap invite ordinary citizens to produce high-quality interactive digital maps. Mitchell's call for tool maps is still relevant, but with such exceptions as proprietary maps, antiques, and some one-of-a-kind representations, most maps can now be seen as both reference and tool maps.

Just as word processors, online dictionaries, dictation, social media, and other tools have made writing text more accessible to many, new cartography tools democratize mapping. And, as is the case for writing, issues of accuracy, precision, evidence base, aesthetics, and overall quality remain. The tools are useful only with intelligent application.

REPRESENTATIONS OF REPRESENTATIONS

The idea of map as workspace takes us from map as faithful representation of physical reality to map as thinking tool. It also opens up the possibility that the cartographer's work is manipulating symbols that themselves represent other symbols.

Think of representing your own house or apartment grounds in a map. You could never portray it with 100 percent accuracy and total precision. You could not access it directly in all of its dimensions. If you wanted to put in the effort, you could measure distances, say, from a building to a street. You could measure angles and north-south orientation. You could take photos of significant features. Soon, you would have a notebook filled with measurements. You might have tables of numbers and specifications.

From that research, you could produce a map, essentially creating spatially distributed elements to show the relations among elements of the system drawing from the photos or numbers in your notebook. You would make choices about what was meaningful to display and what could be omitted. Your work on that map is a thinking process in which you would convert one text (the notebook and photos) into another (the map).

This process would be akin to what anyone must do in working with alphabetic or mathematical text. It means comprehending symbols by making new ones. Maps are, thus, a crucial language for thinking.

THE SOCIAL CREATION OF NATURE

When we think with maps, we are not simply reproducing nature, whether that be a system of rivers, a cityscape, or a collection of data on political preferences. We are creating nature. That does not mean that we do not have some collective common ground for understanding the way things are, but that through a map we express some purpose, some social function applied to the realm in question.

Neil Evernden writes,

> A forest may be a mythical realm or a stock of unused lumber, but either way, it is able to serve a social function. It is, in that sense, never *itself* but always *ours*, our "system" of distinctions among the worldly phenomena . . . it is our habit, and perhaps an inevitable one, to subsequently construe nature as the *source itself*. Yet nature is not the well, but the bucket, and a leaky one at that.[25]

A map of a forest might present it as a mythical realm with talking trees and fantastic creatures. Or, it might represent the resources of the forest—lumber, minerals, food sources—that could be exploited economically. It might highlight the possibilities for recreation, or the threats that pollution and habitat destruction pose. Whatever the case, the map participates in the social creation of the forest, defining it for some purpose of argumentation, problem solving, or aesthetic expression.

MAPS AS WORKSPACES FOR LEARNING

Once we think of maps as workspaces for making sense of the world, we might ask why they should be restricted to spatially distributed elements.

One could, for example, use time to indicate location, as when someone says, "that bridge is a lo-o-o-ng way from here." Many maps use combined spatial and temporal indicators. Think of an online map of a city's history, which can be viewed in an animation mode, showing the growth and other changes in the city.

After extending from space to time, we can imagine maps made of sounds, music, tastes and smells, words and numbers, and other media. This leads to a further revision of our definition: "A map is a workspace in which the user creates and manipulates a representation to make sense of the relations among elements of some system." The emphasis here is on making sense. That means that maps are imbued with intentionality, whether on the part of the creator or of the user. Could a photograph of a scene be a map? It can become one when a person uses it to make sense of the area it depicts. Is a detailed USGS topographic map a map? Only if the person can interpret it or see it as a map. Otherwise, it may be simply a pleasing design with curving, mostly parallel lines. We are not given maps as physical objects but make them in our minds.

This means that a map is not a reproduction of some aspect of the world but a commentary about it. When we look at a street map, we usually want to see street names, for example, "Pamet Point Road." These names are in a font size as large as the portrayal of the road itself, even though we know full well that no street has a gigantic "P," "a," or "m" painted on the road surface or along the verge. Adding those names enables the story that the cartographer tells about the physical space or the one that we intuit from reading it. So, we could say, "A map is a story the author expresses or the reader infers, built upon a spatialized representation of the relations among elements of some system."

Avner Segall writes, "Every map tells a story that serves a purpose and advances an interest. While we tend to think of a map as a neutral depiction of the real world, it is a record that selects, delimits, and inventories only a part of it. The traditional perception of a map as a mirror, a precise graphic representation, is thus untenable. All maps embody their authors' perspectives, assumptions, and biases."[26]

THE IMPULSE FOR LEARNING

The tradition of progressive education saw mapping as a core activity for thinking.[27] Through maps we contextualize and connect ideas. We venture into the larger world and communicate our developing understanding of it. We tell stories about the patterns and relations we discover.

Maps are, thus, tools for learning. And what do learners need? At a minimum they want to enter into the social world through language, to make things, to inquire, and to express feelings and ideas. Dewey saw these impulses, rather than the traditional disciplines, as the foundation for the curriculum, built upon and amplified by the heritage of the disciplines.[28] Maps play an important role for each of these impulses. The educational challenge is to nurture the impulses for lifelong learning.

If we view maps as stories, we can understand better why they serve such diverse purposes. Stories are key to the core activities in all of life: communication, constructing the world around us, inquiry, and expression. By making those stories into spatial representations, we can explore them and the world they describe in greater detail, consider their consequences, and use them for thinking and learning.

NOTES

1. Gary Benenson and James L. Neujahr, *Mapping* (Portsmouth, NH: Heinemann Educational Books, 2002), 20.
2. *Strange Maps: An Atlas of Cartographic Curiosities* (New York: Viking Studio, 2009), xi–xii.
3. Joshua Stevens, Jennifer M. Smith, and Raechel A. Bianchetti, *Mapping Our Changing World*, edited by Alan M. MacEachren and Donna J. Peuquet (University Park: Department of Geography, Pennsylvania State University, 2012).
4. Intergovernmental Committee on Surveying and Mapping, "Types of Maps," *Fundamentals of Mapping* (blog), n.d., http://www.icsm.gov.au.
5. "What President Roosevelt Did to the Map of the U.S. in Four Years with $6,500,000,000," *Life*, January 4, 1937.
6. From Greek *clados* ("branch") and *gramma* ("character").
7. Timothy M. Lenton et al., "Climate Tipping Points—Too Risky to Bet Against," *Nature*, November 27, 2019, 575, 592–95.
8. "Towards a Periodic Table of Visualization Methods of Management," in *Proceedings of the IASTED International Conference on Graphics and Visualization in Engineering*, GVE '07 (Anaheim, CA: ACTA, 2007), 83–88. The authors draw from the work of Edward R. Tufte and others on methods of visualization.
9. Stephen E. Nash, "Lost or Found? A Stick Chart from the Marshall Islands," *Sapiens*, July 25, 2016, https://www.sapiens.org/column/curiosities/stick-chart-marshall-islands/.
10. Rebecca Solnit, *Wanderlust: A History of Walking* (New York: Verso, 2002), 29.
11. Solnit, *Wanderlust*, 276.
12. Simon Greenwold, "Spatial Computing" (master's thesis, Massachusetts Institute of Technology, 2003), 2.

13. S. Chang et al., "Mobility Network Models of COVID-19 Explain Inequities and Inform Reopening," *Nature*, 2020.

14. Madelyn R. Sanfilippo et al., "Disaster Privacy/Privacy Disaster," *Journal of the Association for Information Science and Technology* 71, no. 9 (March 13, 2020): 1002–14.

15. Rebekah Carter, "What Is Spatial Computing? The Basics," *UC Today*, January 27, 2020.

16. Daniel Rosenberg and Anthony Grafton, *Cartographies of Time: A History of the Timeline* (New York: Princeton Architectural Press, 2010), 246.

17. Joseph Graham, William Newman, and John Stacy, "Geologic Time Spiral: A Path to the Past," *United States Geological Survey* (blog), 2008, https://pubs.usgs.gov/gip/2008/58/.

18. Denis Wood, *Rethinking the Power of Maps* (New York: Guilford, 2010).

19. For example, the U.S. Geological Survey topographic maps match up with neighboring maps in eight major compass directions. They even indicate which map is the neighbor. On the downside, however, the National Topographic System Maps of Canada, which touch USGS maps along a three-thousand-mile border, do not align directly with the U.S. maps. Worse than that, in most cases each set shows the topographic details only for its own country.

20. Other specifications, or meta-features, include *scale* (what is the ratio of a unit length on the map to the corresponding one in the world?), *date* (when was the map created or updated?), *cartographer* (who made the map?), *projection* (how a three-dimensional surface is rendered in two dimensions), *purpose* (what was the primary purpose of the map?).

21. The figure of a tree on a map might indicate one tree, a wooded area, an undeveloped zone, or simply be decorative. A road mapmaker might use the term *road* to mean only major highways, whereas another would describe a dirt track as a road. What counts as a road might depend on the scale, with a one-lane dirt track qualifying as a road only for a large-scale map. Similarly, the mapmaker who produces a histogram would need to specify ranges or intervals for some continuous variable so that it could indicate the probability distribution of that variable (i.e., how much fell into each interval).

22. Dewey, "The Child and the Curriculum," MW 2:283–84.

23. Dewey, "The Child and the Curriculum," MW 2:284. Steven Fesmire has written cogently about Dewey and maps. He makes explicit the idea of maps as propositions, or tools for cultural inquiry. A concise but rich summary can be found in his book *Dewey* (New York: Routledge, 2015), especially in the second chapter, "Metaphysics Reconstructed."

24. Lucy Sprague Mitchell, *Young Geographers: How They Explore the World and How They Map the World*, 4th ed. (New York: Bank Street College of Education, 1991), 40.

25. Neil Evernden, *The Social Creation of Nature* (Baltimore, MD: Johns Hopkins University Press, 1992), 110.

26. Avner Segall, "Maps as Stories about the World," *Social Studies and the Young Learner* 16, no. 1 (October 2003): 21.

27. Progressive education did not seek to eliminate formal schooling, or simply to replace it with free, open, discovery learning. Instead, it argued that the primary resource for learning was the desire of the learner to participate fully in life. Tools like textbooks had value but only when they could build upon the concerns and perceptions of the learner. This implies that education maintain a respect for diversity in all its forms. Beyond the individual, it saw the goal of helping learners to become critical, socially engaged citizens.

28. John Dewey, "The School and Society." MW 1:5–109.

Conclusion

The Map Ahead

> Geography is a fantastic subject with extraordinary educational potential for informing future citizens. Geography not only helps pupils understand the savage power and differentiated impact of tsunamis but also the challenges in coping with the aftermath. It challenges pupils with "real world" issues from the local to the global. The subject is rich in multimodal information and communications skills, and can induct pupils into the £20 billion geographic information systems industry that underpins almost all economic activity in the modern world. So why is geography experiencing something of a mini crisis right now?
>
> —David Lambert, "Why Subjects Really Matter"[1]

A book like this cannot be finished. This is not simply because a writer is never satisfied, but because the realm of maps and their uses is unbounded. As soon as one thinks that the topic has been at least glossed over, an entire region can pop up and demand attention.

So, this book is at most a limited approximation, hinting at interesting issues or curious examples. The hope is that it conveys some of what maps are, why they matter, and how they become tools for thought in every imaginable area. In so doing, it argues for a reconceptualization, not just of maps and mapping, but of our models of language and thinking, and of education.

If we accept that maps are, like literacy or mathematics, vital for cognition, then it follows that we should develop better strategies for how we teach and learn. Among these are the following.

REVIVE GEOGRAPHY EDUCATION

The aforementioned epigraph asserts that geography has an "extraordinary educational potential for informing future citizens." Developing citizens are expected to inform national policy or to choose between candidates who will make decisions about, say, war and peace or international treaties in the Middle East. Should they be able to distinguish between Iran and Iraq on a map, or have some idea about historical alliances, languages and religions of different regions, or oil supplies and how oil is transported?

Yet, many people have difficulty using maps and cannot locate trouble spots on a blank map. Concerned commentators decry widespread geographic illiteracy, for example, pointing out how few American high school students can find Brazil on a map.[2]

Moreover, as Lambert notes, geography amplifies diverse multimodal information and communications skills and carries enormous economic value. These are more than enough reasons to include geography in any well-balanced curriculum. However, these arguments for the subject matter of geography are only a small part of the argument of this book.

PLACE GEOGRAPHIC THINKING AT THE CENTER

Certainly, acquiring specific map skills and knowledge can be useful. We could recognize the way that geography both calls upon and extends multi-modal information and communications skills. However, the argument here is about what a wide-ranging use of maps implies for education in general, not just about geography education, as important as that may be.

The examples in this book show many ways that maps, broadly conceived, can be useful in every realm of inquiry, for diverse modes of communication, and across an unlimited range of expression. They offer models and insights that can be useful for individuals in any of these situations and for educators seeking ideas to employ or adapt.

Learning through maps, both reading and constructing them, is essential to thinking and interpreting across all of the disciplines. We should appreciate geography as a unifying discipline for the entire curriculum. It is crucial for democratic education because democracy requires understanding our place in the world, our relation to others, and the employment of all of our cognitive capacity.

ENGAGE WITH THE FULL ARRAY OF MAPPING TOOLS

It is impossible to name a discipline that is not changing through the use of new spatialization and visualization tools, for instance, data analytics, virtual reality, immersive spaces, geographic information systems, satellite images, and multimodal communications. Educators should consider how to help students engage with these tools, both as competent users and as critics. Our knowledge of a discipline remains partial until we understand how it makes use of multiple media for representing and manipulating spatially defined elements.

Most of these tools are highly accessible, or at least have versions offering open access and easy-to-use online interfaces. Meanwhile, such traditional mapping tools as atlases, maps, globes, diagrams, hand-drawn maps, and more are still useful and need to be conceived in relation to the new digital tools. The diversity of mapping tools is one of its primary contributions to a robust and healthy ecosystem for learning.

EMBED MAPPING WITHIN THE NETWORK OF OTHER LEARNING TOOLS

Rugg's curriculum from the 1930s showed how maps could be viewed—not as isolated resources containing specific information, but as part of a network of learning tools, including texts, tables of data and, of course, critical discussion. In the effort to revive geography, or some modern incarnation, for example, geographic information systems, we need to guard against the temptation to ensconce it as a separate item, one more thing to be learned. Instead, it should be viewed as part of a larger network for learning.

RECOGNIZE THE EMERGENT PROPERTIES OF MAPS

Maps need to be understood as more than static repositories of information. They can be that, of course, often holding far more information than one imagines, as in the case of the *National Geographic* globe discussed earlier. However, their greatest power may come when, especially in concert with other learning tools, they become sites for exploration and discovery of ideas not explicitly represented in the map.

A simple example is the use of layers in geographic information systems. One might discover the sources of pollution in an area based upon the representation of factories and housing, plant and animal life, agriculture, measurements of water quality, and other such information. Going further, one might link that discovery with historical data showing how the same area developed throughout time.

BECOME WRITERS OF MAPS, NOT JUST READERS

The diverse types and uses of maps fill a sumptuous archive offering paths to understanding in every discipline. They enlarge the knowledge we might gain from reading or calculation. For virtually any form of inquiry, we can turn to maps, whether in digital or physical form, at any level of detail, on any topic. These are what Lucy Sprague Mitchell calls *reference maps*, an invaluable tool for learning.

Mitchell also identified and enacted through the Bank Street College curriculum a second type, called *tool maps.*[3] These may appear cruder, less accurate, or incomplete, but they are equally important for thinking. Tool maps are made in the process of inquiry. They represent the inquirer's developing conception of a question. In contrast to a reference map, a tool map is less for communication to a general public and more for the inquirer to work out ideas—especially to situate their own experience within the topic.

The tool mapmaker is a writer of maps. We know that writing text is crucial for learning.[4] Similarly, writing maps can be as valuable for acquiring knowledge, if not more so, than reading maps. Moreover, when students learn to create maps, they become more critical consumers of maps that others make.

APPLY CRITICAL INTERPRETATION TO MAPS

Any map distorts the reality it purports to represent. Sometimes distortions are deliberate attempts to mislead but, in general, these distortions are due to necessary attempts to simplify for clarity or to apply assumptions about data and potential users.

All too often, a student is guided to look up something on a map as if that would provide the complete and final truth. A more useful approach is to examine the rhetoric of a map just as one would that of an essay: When was it written? By whom? What was the audience? What assumptions does it make? What sources does it reference? And so on.

A simple example concerns how one might judge the age of a map or globe. A copyright date might help, but that in itself is not always fully reliable. Another approach is to use various tables of data showing when political events occurred that might be reflected on a map.[5] For instance, in 1970 Muscat and Oman changed its name to Oman. A map using the older name would have been created before 1970, whereas one with Oman alone would likely be from 1970 or later.

SEEK A SENSE OF THE WHOLE

The very power of the interactive, digital map—its ability to shift scale or to jump from Kankakee to Kathmandu—also makes it easy to get lost, the one thing many of us expect a map to prevent. This is why many people are lost when they are out of cell range and the directions assistant goes silent, even if they can still see the map. We have come to rely on step-by-step directions, having lost a *sense of the whole*.

That sense of the whole is important for democratic education, both for its ultimate purpose and for its enactment in schools. We need an integrated understanding that extends well beyond learning specific facts and skills, whether those be in history, science, language, or geography itself.

The geographic imagination defined in the introduction is essentially a recognition of what is needed to regain a sense of the whole. It draws upon geography in its broad sense as a discipline concerned with both the social and the physical worlds; with appreciating and understanding other living things and, of course, the humans dwelling there; and with helping us place ourselves in those worlds.

Nurturing that imagination is not simply a matter of a technology fix, nor is it a call to return to, say, the *National Geographic* political globe and geometer, much less to the quadrivium as the core curriculum. Instead, it is to recognize the need for what Lucy Sprague Mitchell call *map thinking*. That, in turn, calls for a laboratory approach to geography, seeing it, and individual maps, as tools for problem solving and making sense of ourselves and our world.

PRACTICE DEMOCRATIC EDUCATION

Learning is ecosystem-based, and learners develop primarily through direct interaction in their own ecosystem, whether that includes home, family, friendships, school, work, hobbies, or other activities. Accordingly, we need to consider how to ensure a healthy learning ecosystem, one that nurtures healthy growth for individuals and society: How can we guide the educational enterprise by an ethical vision, not simply a technocratic one of transmitting isolated facts and skills? How can we promote the development of critical, socially engaged citizens? How can we see diversity in all its forms as a resource for learning? What materials and activities best promote *eudaimonia* (flourishing) for all?

This book frames that as *democratic education*. This is an educational ideal in which democracy is both a goal and a method of instruction. As a goal, democratic education adds to individual growth the idea that a key purpose of education is to foster a society in which all people are valued, regardless of nationality, race, religion, language, sexuality, or physical ability, and that they have full opportunity for their own participation and growth. Democratic education is, thus, growth-oriented, progressive, and humanistic.

Moreover, through the process of participatory education, citizens develop an enlarged capacity to critique, participate in, and sustain a flourishing society. This inevitably means that education must be integrated across subject areas, time, and between school and community.

As a method, democratic education includes self-determination within a community, as well as such values as social justice, understanding across differences, respect, and trust. Thus, it implies student-centered learning and open processes for learning. It also implies learning about the physical and social world, which we create and inhabit, and being engaged with that world. It is, thus, inextricable from the best sense of geography education.

More important, the wide use of maps shows how education can be changed to move toward that ideal of democratic education. Maps, and spatialization in general, are a fundamental way of augmenting our thinking, memory, and communication. Viewed as tools for thinking, they can help us open up the processes of education and make it more democratic in both aim and method.

FINDING OUR WAY

This book is a guide, a map one might say, to the universe of mapping, and of maps, their multitudinous forms and uses, their diverse users, and their fascinating history and cultural associations. If it has succeeded, even in part, it should be obvious that what it shows is only a tiny glimpse of that universe.

The more we see the incredible variety of maps that people create and the purposes they use them for, the more we see that maps are simply a fundamental aspect of thinking. Although the book has attempted to define maps, any definition ultimately fails unless we can find a way to put a limit on human thought.

Nietzsche saw the problem we face in all areas of language use: every word—map, mapping, cartography, geography, path, boundary, territory, and so forth—immediately becomes a concept referring to the experience that spawned it but also called upon to fit innumerable, more or less similar, but necessarily unequal cases. Thus, "every concept originates through our equating what is unequal."[6] As Nietzsche explains, this process of category formation removes us from the "unique and wholly individualized original experience" that just occurred. Thus, every concept takes us away from reality.

Rather than seeking unassailable definitions, or attempting to impose order on the unruly universe of maps, it may be best to think of them in terms of how people use those terms, what mapmakers do, and how all of us use maps. Our journey has shown that we have made maps since prehistoric times, longer probably than writing or mathematics, possibly longer than all but the most rudimentary speech. They are essential to our thinking and communicating in virtually every realm.

It follows that maps for learning are much more than visual aids. As models for the world that we seek to understand, they are the arena in which we articulate what we are learning, try out new ideas, consider what-if scenarios, face challenges, and communicate with others. They aid our exploration and consolidate our knowledge. They are the best way we have to picture what is happening inside us as we extract meaning from experience and learn how to live more fulfilling lives.

NOTES

1. David Lambert, "Why Subjects Really Matter: A Personal View," *GTTP Think-Piece* (blog), 2018, https://www.geography.org.uk/write/MediaUploads/Teacher%20 education/GA_PRGTIPlambertwhy.pdf.
2. Phil Gersmehl, *Teaching Geography*, 3rd ed. (New York: Guilford, 2014); David Lambert and David Balderstone, *Learning to Teach Geography in the Secondary School*, 2nd ed. (New York: Routledge, 2010); Gary Benenson and James L. Neujahr, *Mapping* (Portsmouth, NH: Heinemann Educational Books, 2002).
3. Lucy Sprague Mitchell, *Young Geographers: How They Explore the World and How They Map the World*, 4th ed. (New York: Bank Street College of Education, 1991).
4. Robert P. Parker and Vera Goodkin, *The Consequences of Writing: Enhancing Learning in the Disciplines* (Upper Montclair, NJ: Boynton/Cook, 1987).
5. For example, http://www.longwood.edu/cleanva/datingglobes.htm.
6. Friedrich Nietzsche, "On Truth and Lie in an Extra-Moral Sense," in *The Portable Nietzsche*, edited by Walter Kaufmann (New York: Viking, 1976), 42.

Index

Boldface numbers indicate the page defining a term. Italic numbers indicate reference to a figure.

About the Author

Bertram C. Bruce is professor emeritus of information science at the University of Illinois. His work spans artificial intelligence, literacy, math and science education, technology-enhanced learning, inquiry-based learning, teacher learning, and collaborative community-based projects. Throughout he has emphasized democratic education, asking, "How can we guide the educational enterprise by an ethical vision, not simply a technocratic one of transmitting isolated facts and skills?" His latest book is *Education's Ecosystems: Learning through Life* (Rowman & Littlefield, 2020).

www.ingramcontent.com/pod-product-compliance
Ingram Content Group UK Ltd.
Pitfield, Milton Keynes, MK11 3LW, UK
UKHW012040150926
475560UK00001B/101

9 781475 859287